Tin-Based Antitumour Drugs

NATO ASI Series

Advanced Science Institutes Series

A series presenting the results of activities sponsored by the NATO Science Committee, which aims at the dissemination of advanced scientific and technological knowledge, with a view to strengthening links between scientific communities.

The Series is published by an international board of publishers in conjunction with the NATO Scientific Affairs Division

A Life Sciences B Physics	Plenum Publishing Corporation London and New York

C Mathematical and
  Physical Sciences
D Behavioural and
  Social Sciences
E Applied Sciences
	Kluwer Academic Publishers
	Dordrecht, Boston and London

F Computer and Systems Sciences G Ecological Sciences H Cell Biology	Springer-Verlag Berlin Heidelberg New York London Paris Tokyo Hong Kong

Series H: Cell Biology Vol. 37

Tin-Based Antitumour Drugs

Edited by
M. Gielen
Vrije Universiteit Brussel, Room 8G512, Pleinlaan 2,
B-1050 Brussels, Belgium

Springer-Verlag Berlin Heidelberg New York
London Paris Tokyo Hong Kong
Published in cooperation with NATO Scientific Affairs Division

Proceedings of the NATO Advanced Research Workshop on The Effect of Tin upon Malignant Cell Growth held in Brussels, Belgium, July 16–20, 1989

ISBN 3-540-50417-6 Springer-Verlag Berlin Heidelberg New York
ISBN 0-387-50417-6 Springer-Verlag New York Berlin Heidelberg

Library of Congress Cataloging-in-Publication Data. NATO Advanced Research Workshop on the Effect of Tin upon Malignant Cell Growth (1989 : Brussels, Belgium) Tin-based antitumour drugs / edited by M. Gielen. p. cm.—(NATO ASI series. Series H, Cell biology ; vol. 37) "Proceedings of the NATO Advanced Research Workshop on the Effect of Tin upon Malignant Cell Growth, held in Brussels, Belgium, July 16–20, 1989"—T.p. verso. Includes bibliographical references.
ISBN 0-387-50417-6 (U.S. : alk. paper)
1. Organotin compounds—Therapeutic use—Testing—Congresses. 2. Antineoplastic agents—Testing—Congresses. I. Gielen, M. (Marcel), 1938– . II. Title. III. Series. [DNLM: 1. Antineoplastic Agents—therapeutic use—congresses. 2. Organotin Compounds—therapeutic use—congresses. QZ 267 N2785t] RC271.073N37 1990 616.99'4061—dc20   DNLM/DLC   for Library of Congress   89-26162

This work is subject to copyright. All rights are reserved, whether the whole or part of the material is concerned, specifically the rights of translation, reprinting, re-use of illustrations, recitation, broadcasting, reproduction on microfilms or in other ways, and storage in data banks. Duplication of this publication or parts thereof is only permitted under the provisions of the German Copyright Law of September 9, 1965, in its version of June 24, 1985, and a copyright fee must always be paid. Violations fall under the prosecution act of the German Copyright Law.

© Springer-Verlag Berlin Heidelberg 1990
Printed in Germany

Printing: Druckhaus Beltz, Hemsbach; Binding: J. Schäffer GmbH & Co. KG, Grünstadt
2131/3140-543210 – Printed on acid-free-paper

FOREWORD

Whereas platinum compounds are already clinically used as anticancer agents, tin compounds exhibiting high enough antitumour activity have not yet been found to enter the clinical phase. This is paradoxical with the observation that 50% of the tested compounds showed some activity, which is an abnormally high percentage. This is probably due to the fact that research of this type with tin compounds started more recently than with platinum compounds and that almost exclusively known compounds coming from laboratories working in the field of organotin compounds have been blindly tested.

Some research directed towards the development of organotin compounds with some biologically favourable properties, for instance derivatives with hydrophilic or lipophilic, or with increased bio-availability characteristics, has started only recently.

We thought that was now the appropriate time to bring together experts from different disciplines (biochemists, pharmacologists, organotin chemists, oncologists, etc.) working in the field of tin-based antitumour drugs, to contribute to the critical assessment of existing knowledge in this new important topic, to identify the directions for future research and to promote close working relations between the different countries and professional experiences gathered in this workshop.

I think that this meeting was a fruitful one, not only because of high standard review papers, but also because of the contacts that have been possible between the participants through the many discussions that have been included in the program, through the social activities, the visit to Brussels downtown, the banquet, and through the coffee breaks and the lunches that were organised on the congress site.

I am glad that so many participants attended this NATO Advanced Research Workshop, representing so many countries, that several participants to the former meetings of this series were again attending this one, and that a lot of newcomers were also present, like for instance the numerous Turkish delegation.

I would like to thank my coworkers, Ann Delmotte, Rudi Willem, Nadia Cornand and Guy Nuttin for their precious help in organising this meeting, and NATO for their financial help.

Marcel Gielen
Director of the NATO ARW

CONTENTS

The Role of Non-Platinum Complexes in Cancer Therapy   1
Bernhard K. Keppler

Tumor-Inhibiting Metal Complexes and the Development of Cisplatin	1
The Development of New Tumor-Inhibiting Metal Complexes	9
Fundamental Strategies	9
Methods	9
Synthesis of Direct Derivatives of Cisplatin	15
Basic Ideas	15
The Development of Non-Platinum Complexes	17
Non-Platinum Complexes in Preclinical Trials	17
Tumor-inhibiting Ruthenium Complexes	24
Non-Platinum Complexes in Clinical Studies	39
Gallium and Germanium Compounds in Clinical Studies	40
Preclinical and Clinical Development of Budotitane	41
Structure-Activity Relation of Tumor-Inhibiting Bis(β-diketonato) Metal Complexes	43
Variation of the (β-Diketonato) Ligand	44
Variation of the Central Metal	47
Variation of the Group X	48
Antitumor Activity in Other Transplantable Tumor Models	50
Therapy Results on Autochtonous, AMMN-Induced, Colorectal Tumors with Budotitane	51
Toxicity of Budotitane	53
Budotitane Clinical Phase I Study	54
Drug Targeting	55
Perspectives	58
References	60

Tin Compounds and their Potential as Pharmaceutical Agents   69
Alan Crowe

Introduction	69
Tin Protoporphyrin for the Treatment of Neonatal Jaundice	72
Antitumour Properties of Tin Chemicals	80
Mode of Action	91
The Use of Tin Derivatives in the Photodynamic Therapy of Cancer	95
Tin Derivatives as Antiviral Agents	102
Other Pharmaceutical Uses	104
Conclusion	105
Acknowledgements	108
References	108

The Role of Natural Tin Hormones in Senescence: a Hypothesis   115
Nate F. Cardarelli

Background: Immune Involvement	115
Background: Geroprotective Effects of Tin	120
Capitulation	121
Ontogeny of the Immune System	121
Immune Theory of Aging	123
Thymus-Pineal Axis	124
Life Pattern Hypothesis	127
Tin Deprivation and Life Span	129
Tin and the Life Pattern	131
Summary	136
References	139

The Speciation and Bioavailability of Tin in Biofluids   147
J.R. Duffield, C.R. Morris, D.M. Morrish, J.A. Vesey and David R. Williams

Introduction	147
Historical Outline	147
Speciation Studies	149
Experimental	150
Materials	150
Procedure	150
Results	153
Formation Constants Measurement	153
Regression Analysis	155
Speciation Modelling	155
Discussion	158
Tin in Canned Food	158
Bioavailability	160
Toxicity and Essentiality	161
Conclusions	163
Acknowledgements	164
References	164

Cellular Interactions of Organotin Compounds in Relation to their Antitumor Activity   169
André Penninks

Introduction	169
Direct Cytotoxic Effects	170
Disturbances of Mitochondrial Energetics *in vitro*	173

Disturbances of Cellular Energetics *in vitro*   176
Disturbances of Macromolecular Synthesis *in vitro*   178
Summary and Conclusions on the Cellular Effects of Organotins   180
Vivo-Vitro Relation of Antitumour Activity of Organotins   182
Mode of Action of Organotins in Relation to their Antitumor Activity   184
References   186

Selectivity of Antiproliferative Effects of Dialkyltin Compounds *in vitro* and *in vivo*   191
Gerhard Hennighausen and Stanislaw Szymaniec

Computer Assisted Structure-Activity Correlations of Organotin Compounds as Potential Anticancer and Anti-HIV Agents   201
Ven Narayanan, Mohamed Nasr and Kenneth D. Paull

Introduction   201
Results   201
Discussion   203
Conclusions   216
References   216

Route of Administration is a Determinant of the Tissue Disposition and Effects of TBTO on Cytochrome P-450-Dependent Drug Metabolism   219
Daniel W. Rosenberg

Introduction   219
Materials and Methods   219
Results and Discussion   221
Acknowledgements   224
References   225

THE ROLE OF NON-PLATINUM COMPLEXES IN CANCER THERAPY

B.K. Keppler
Anorganisch-Chemisches Institut der Universität Heidelberg,
Im Neuenheimer Feld 270, 6900 Heidelberg, FRG

TUMOR-INHIBITING METAL COMPLEXES AND THE DEVELOPMENT OF CISPLATIN

Metal complexes as pharmaceutical agents have been used since early history, but therapeutic efficacy in today's meaning of the term was first confirmed on the basis of the examples of salvarsan (1910), particularly efficient in cases of syphilis, and some organic mercury compounds, such as novasurol (1919), and salyrgan (1924), which were used as diuretic agents. These drugs have gradually come to be replaced by compounds from organic chemistry that exhibit better activity. Nowadays drugs from inorganic chemistry are mainly represented by auranofin (INN), (2,3,4,6-tetra-O-acetyl-1-thio-1-ß-D-glucopyranosato)(triethylphosphine)gold(I) (Fig. 1), active against primary chronic polyarthritis (PCP) (Berners-Price and Sadler, 1985; Lewis and Walz, 1982), sodium nitroprusside, nipruss^R, disodiumpentacyanonitrosylferrate(II)dihydrate, $Na_2[Fe(NO)(CN)_5] \times 2H_2O$, used as an emergency drug in the case of high blood pressure crises, lithium salts, used in psychiatry (Pöldinger, 1982), many preparations for local application in dermatology and gastroenterology, and metal salts for the prevention of deficiencies. In cancer therapy, the only drug from inorganic chemistry to be under routine clinical use is cisplatin (INN), cis-diamminedichloroplatinum(II) (Fig. 2). This drug was synthesized for the first time by Michele Peyrone and was published in 1844 in the "Annals of Chemistry and Pharmacy" (Peyrone, 1844).

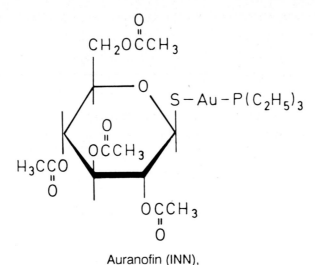

Auranofin (INN),
(2,3,4,6-Tetra-O-acetyl-1-thio-1-ß-D-
glucopyranosato)(triethylphosphine)gold(I)

Fig. 1. Auranofin (INN).

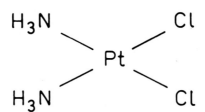

Cisplatin(INN), cis-Diamminedichloroplatinum(II)

Fig. 2. Cisplatin (INN), cis-diamminedichloroplatinum(II), widely used as anticancer agent in the clinic today.

More than a century later, in 1969, Barnett Rosenberg discovered the tumor-inhibiting qualities of cisplatin. Rosenberg, while examining the influence of an electrical field on bacteria growth, found filament growth of *escherichia coli* bacteria. This was caused by the experimental conditions which favored the forming of cis-configurated platinum complexes.

The selective influence on cell division, along with an unrestrained growth, led Rosenberg to realize that compounds such as these might also be capable of inhibiting tumor growth. He synthesized some simple platinum amine complexes and, screening them in the sarcoma 180 and in the murine L 1210 leukemia, achieved a reduction of tumor weight and a prolongation of survival time of tumor-bearing animals using cis-$Pt(NH_3)_2Cl_4$, cis-$Pt(NH_3)_2Cl_2$, $PtenCl_2$, and $PtenCl_4$ (Fig. 3) (Rosenberg and VanCamp, 1969, 1970; Rosenberg, 1975, 1978; Kociba et al., 1970). He proved that all these cis-configurated compounds had an outstanding tumor-inhibiting effect in animal experiments. In contrast to this, the corresponding trans- compounds were found to be inactive.

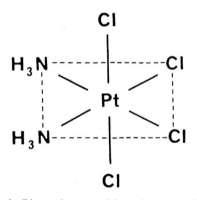
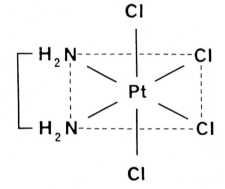

Fig. 3. Rosenberg was the first to examine these platinum complexes with regard to tumor-inhibiting properties. Cis-Diamminetetrachloroplatinum(IV) was the substance that caused filament growth of escherichia coli bacteria in his experiments.

At first there were considerable reservations regarding these heavy metal compounds. Nevertheless, cisplatin entered clinical studies and soon showed amazingly good effects on various advanced tumors. Today, cisplatin is frequently used in combination with other antitumor agents. It is effective mainly against testicular carcinomas, ovarian carcinomas, bladder tumors, and tumors of the head and neck.

Testicular carcinomas, which had almost invariably caused death before cisplatin was discovered, has transformed into a curable disease in most cases today, thanks to the use of cisplatin. This effect is illustrated by the lung x-rays of a patient suffering from testicular carcinoma (Fig. 4). Here, the impressive remission of massive lung metastases after cisplatin administration is evident.

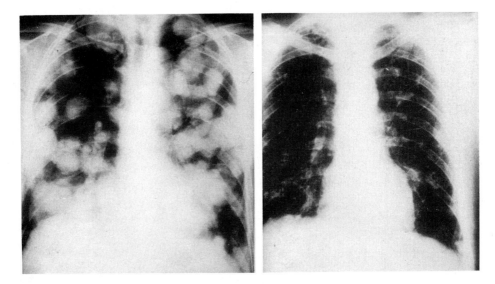

Fig. 4. Lung metastases of a 25-year-old testicular cancer patient; on the left, before, and on the right, after four cycles of cisplatin-containing chemotherapy (Photos: R. Hartenstein, Munich-Harlaching, FRG).

Cisplatin is highly active against some relatively rare tumors, but it does not show any, or only little, effect on tumors that are very common and that account for the major share of cancer mortality today, such as lung tumors and adenotumors of the gastrointestinal tract. The excellent activity of cisplatin against testicular carcinomas has proved that it is possible in principle to find new drugs - in this case from inorganic chemistry - that are capable of curing specific types of tumors. This has of course been a considerable impetus for the inorganic chemist to search for new metal complexes which have similar good activity, but against those types of tumor that are responsible for the major share of cancer mortality today, e.g., adenotumors of the gastrointestine.

The most important side-effect of cisplatin is nephrotoxicity. This used to be dose-limiting, but it can be repressed today by fractionated therapy schemes and forced diuresis with hyperhydration - mostly with mannitol solutions. The extent of this repression is so great that the normally mild cisplatin myelotoxicity can turn into a dose-limiting factor on account of the higher doses which can be applied. Ototoxicity is found less frequently and is manifested, above all, in a hearing loss for high frequencies > 4000 Hz. Peripheral neuropathies are contracted by few patients. Nausea and vomiting are subjectively considered by patients to be the most severe side-effects, but it is precisely this effect that is particularly pronounced with cisplatin therapy, more than with most other antitumor agents. Owing to results of animal experiments it is assumed that cisplatin has carcinogenic effects. The extent of this carcinogenicity in humans is not completely known yet (Kempf and Ivankovic, 1986; Leopold et al. 1979).

As to the mechanism of action of cisplatin, scientists have as yet carried out numerous examinations. The adducts that cisplatin forms with DNA are investigated in particular. It is supposed that these adducts inhibit DNA replication and thus also affect the tumor. Several mechanisms are being discussed in this connection (Fig. 5).

Cisplatin can form intrastrand and interstrand crosslinks between DNA bases. Also, DNA crosslinks with other nitrogen-containing molecules such as proteins are possible, but these are presumably of no importance (Fig. 5).

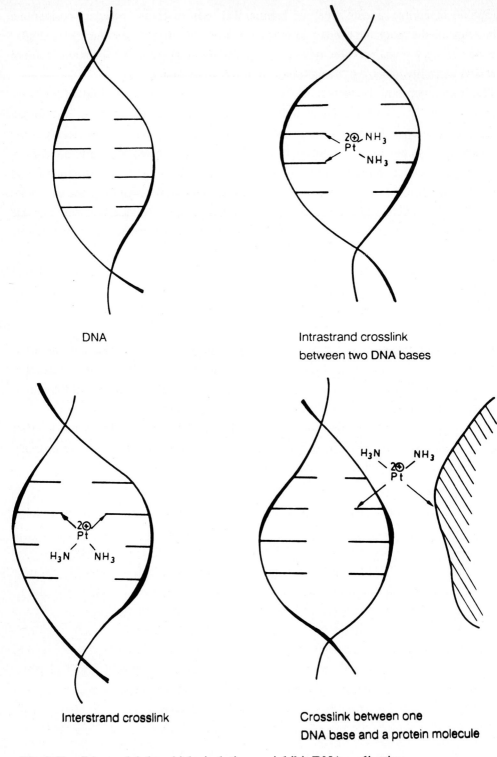

Fig. 5. Possible models by which cisplatin may inhibit DNA replication.

The most frequent adduct is the intrastrand crosslink between neighboring guanine molecules with complexation at N(7) (Fig. 6). Numerous studies regarding this problem have been carried out, among these are x-ray structure analyses with model nucleobases and oligonucleotides (Sherman and Lippard, 1987; Reedijk, 1987; Pinto and Lippard, 1985; Lippard, 1981; Sherman et al. 1985).

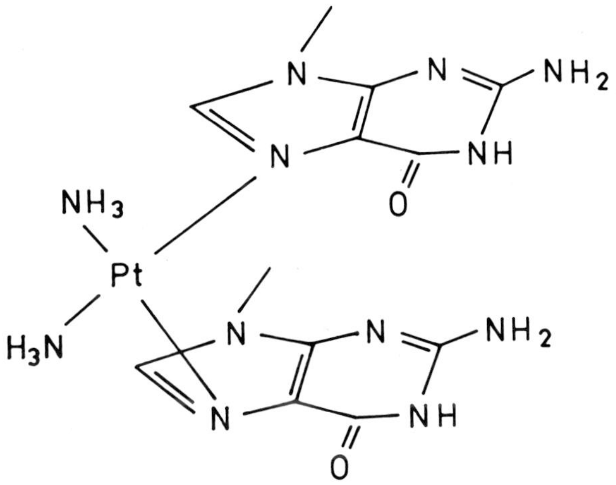

Fig. 6. The intrastrand crosslink between two guanine bases at N(7) is the most frequent adduct cisplatin forms with DNA.

It is not easy to understand how the unspecific DNA adducts shown in Fig. 5 can control an effect of cisplatin as specific as this. On the basis of the mechanisms suggested, cisplatin should in general inhibit rapidly proliferating tissue, that is, it should show what might be called a classic "cytostatic" effect, without exerting a specific influence on certain tumors or certain tissues. As mentioned before, cisplatin is an excellent drug specifically against testicular carcinomas and a few other tumors, while the major part of those types of cancer that account for the major share of cancer mortality is not or only a little sensitive to cisplatin. One might suppose that the reason for this could be a selective accumulation in the relevant tissue. However, it could be demonstrated that there is no strict parallelism between tissue concentration and effectiveness of cisplatin in certain organs. For example, relatively high cisplatin concentrations are found in the skin and in the liver while there is no sufficient activity against tumors of corresponding localisation (von Heyden et al. 1980; Litterst et al. 1979; Ehninger et al. 1984; Prestayko, 1981). Thus further investigation into the mechanism of action of cisplatin is necessary to obtain an overall view of the problem which might explain organ and tumor specifity in its therapeutic effect.

THE DEVELOPMENT OF NEW TUMOR-INHIBITING METAL COMPLEXES

FUNDAMENTAL STRATEGIES

The development of new tumor-inhibiting metal complexes is basically characterized by the three following proceedings (Keppler, 1987):
- *synthesis and activity screen of "direct" cisplatin derivatives*
- *trials with new metal complexes that do not have platinum as their central metal*
- *linking of cancerotoxic platinum compounds or of other tumor-inhibiting metal complexes with carrier molecules or carrier systems in order to achieve an accumulation in certain tissues.*

The first of these three strategies is not very promising in view of the aim of synthesizing substances which have a different spectrum of indication than cisplatin, because experience has shown that the development of "direct" derivatives will lead to substances that are not very different in therapeutic efficacy from the parent compound, because their mechanism of action is similar. The procedure is rather more justified in attempts to decrease toxicity or increase selectivity in comparison with cisplatin. Tumor-inhibiting non-platinum compounds, on the other hand, should be more likely to act on tumors other than those affected by cisplatin, owing to the change in chemical properties in comparison with platinum. Of course it will be much more difficult in this field to find a compound that has any tumor-inhibiting effects at all, because it is not possible to proceed from an established structure-activity relation, as can be done with platinum compounds.

The third possibility, i.e., the linking of cancerotoxic platinum compounds to carrier molecules, will be illustrated on the basis of hormone-linked platinum derivatives, which have an affinity to hormone receptor-positive tumors, and on the basis of osteotropic platinum compounds.

METHODS

There has been much controversial discussion on whether different experimental tumor models are useful for predicting antineoplastic activity in humans. In Heidelberg a procedure has been developed which is based on the combination of transplantable and autochthonous tumor models. Investigations are begun by selecting the compounds on the basis of activity in various transplantable tumor models. These models are chosen in such a way that most of the compounds from the new types of substances to be examined show medium activity in each relevant model, with the result that compounds with major or minor activity can clearly be distinguished. When the chemical structure has been optimized, few candidates with outstanding activity will be further investigated in autochthonous tumor models, which have a higher predictivity for the clinical situation than other models. The general procedure is outlined in Fig. 7 and will be explained more in detail in the following.

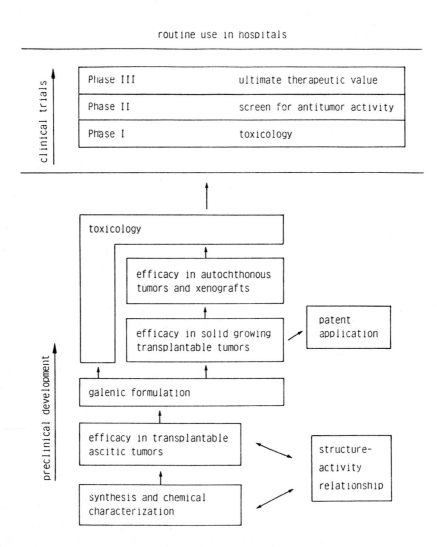

Fig. 7. Preclinical and clinical development of new antitumor agents.

A prerequisite of the screening the biological activity of a substance is, of course, synthesis and faultless chemical characterization. This is followed by the screening of activity in what is known as transplantable ascitic tumors. This is an experimental procedure where the tumor grows intraperitoneally, i.e., in the peritoneal cavity of the animal, usually the mouse, in a liquid, cell-culture-like, form. Therapy is carried out at least 24 hours after tumor transplantation. The potentially active substances are also administered intraperitoneally (Fig. 8). This implies that <u>local</u> activity of the drug at the site of the tumor can be measured. There are various therapy schemes, e.g., on day 1 or on days 1, 2, and 3; on days 1 to 9; or on days 1, 5, and 9 after tumor transplantation (day 0). Numerous tumor models are eligible to this experimental procedure, e.g., leukemia P 388, leukemia L 1210, sarcoma 180, Yoshida sarcoma, Walker 256 carcinosarcoma, DS sarcoma, Stockholm ascitic tumor, Lewis Lung tumor, melanoma B 16, and Ehrlich tumor, among others. In these models, prolongation of survival time of the treated animals is measured in comparison to controls. The so-called T/C value is calculated as follows:

$$T/C\ (\%) = \frac{\text{median survival time of treated animals}}{\text{median survival time of control animals}} \times 100$$

Thus high T/C values indicate good antitumor activity. T/C values should exceed 125 %. These ip/ip models are very sensitive and soon give indication of tumor-inhibiting activity. It is, of course, reasonable to choose the tumor model in such a way that the type of substance under examination will show a medium level of activity. When systematically varying the chemical structure, one may then easily recognize improvement or deterioration in activity. This is not the case if too sensitive a tumor model is chosen where a large number of individual substances from the particular group will reach high T/C values, or if a tumor model is chosen which is insensitive to the particular group of substances and which does not indicate any increase in activity, not even when structural improvements have been made. Thus it is obvious that these first screenings are not designed to cure tumor-bearing animals or to achieve the highest possible survival rates or tumor remission rates but to find the optimum substance in a group of antitumor agents, which may then be examined further.

In the ip/ip models, we only use complexes which can be dissolved undecomposed in a watery solution, or we add solubilizers that are in use for galenic purposes to obtain a clear watery solution. The application of suspensions of insoluble compounds should be avoided because of the artificially high antitumor activity that will be produced in these models this way.

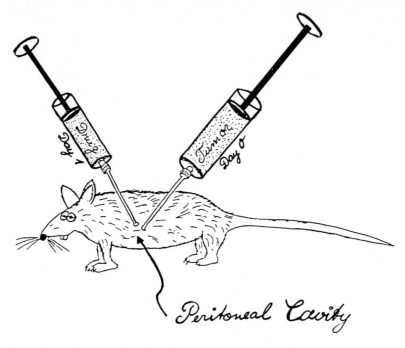

Fig. 8. Experimental procedure in what is known as an ip/ip tumor model. The tumor is transplanted intraperitoneally on day 0 in the form of a cell suspension. Intraperitoneal therapy follows on one or several subsequent days.

When a structure-activity relation has thus been established and a particular substance from a class of compounds has been selected for its promising activity, <u>systemic</u> activity of this complex must be screened. This is done by using solid transplantable tumors, which grow intramuscularly or subcutaneously in the form of tumor nodes. In order to create realistic conditions, the active substance will always be injected intravenously, something which requires that a suitable galenic formulation of the complex be produced beforehand (see Fig. 7). Fig. 9 shows an experimental procedure used in screening systemic activity. Here, the same tumors as listed above are used, but instead of applying the tumor intraperitoneally we will transplant cell suspensions or even tumor fragments intramuscularly or subcutaneously. In most cases tumor weight or tumor volume will be chosen as parameters for the evaluation of the experiment. T/C values are calculated accordingly:

$$T/C\ (\%) = \frac{\text{median tumor weight of treated animals}}{\text{median tumor weight of control animals}} \times 100$$

In this case, low T/C values mean good antitumor activity. T/C values of promising substances should fall below 45 %.

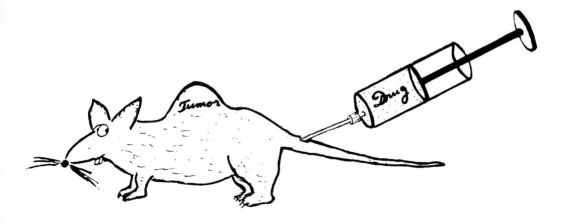

Fig. 9. Experimental procedure in treating a solid tumor. The tumor in this case grows subcutaneously, and the substance is injected intravenously into the tail vein.

When a particularly efficient drug has been selected, evaluation in autochthonous tumors or xenografts will follow (see Fig. 7). These tumor systems are highly sophisticated, and it is at this juncture that cooperation with expert institutes finally becomes necessary. In Heidelberg autochthonous tumor systems are mainly used. What is known as xenografts, i.e., human tumors that are transplanted to immunodeficient mice (nude mice), are also suitable here, but only if the substances in question are investigated in a representative panel rather than in few individual lines.

The autochthonously growing cancer models have been developed during the last 25 years. These test systems mimic the human situation quite closely because of the following characteristics:

1) Tumor origin, -growth and -therapy take place in the same individual.
2) There is genuine tumor histology.
3) Tumor/host interaction is identical to that in humans.
4) The proliferating fraction of tumor cells is lower than in transplanted tumors.
5) Tumor growth is orthotopical, i.e., autochthonous tumors do not grow at artificial sites (e.g., subcutaneously).
6) Autochthonous tumors show a lower sensitivity to conventional cytostatic agents than transplanted tumors (Zeller and Berger, 1984; Schmähl and Berger, 1988).

Since spontaneously occurring autochthonous tumors generally develop too seldom for practical purposes and too late in an animal's lifetime, tumors are chemically induced in several organs. Models which have been developed in this connection include chemically induced leukemias, mammary carcinomas, and colorectal carcinomas. Their general use, however, is restricted to few research centers only, since they are known as relatively "cumbersome" models, which are less suited for a primary screening of high numbers of new, unknown agents. It is hoped, on the other hand, that these models can predict the anticancer efficacy of compounds under investigation much better, since they mimic the human situation more closely than transplanted tumors, including human tumor xenografts grown in nude mice. During the past decade special attention has been paid to the development and characterization of chemically induced colorectal tumors. From various effective carcinogens acetoxymethylmethylnitrosamine was chosen for induction. Treatment usually starts after endoscopical diagnosis of tumor growth which represents the concept of treating established tumors. In this case endoscopy has to be performed in regular time intervals starting after the end of the induction period until the last animal has been found tumor-positive (Berger et al. 1986). This model has been used with success in the development of e.g., the tumor-inhibiting titanium and ruthenium complexes, as will be described later.

Another important aspect of preclinical development is toxicology (Fig. 7). Only a fairly favourable ratio of activity to side-effects of a new drug will render subsequent clinical development possible. However, there is almost always a good correlation between preclinical findings and clinical results as far as toxicity of tumor-inhibiting compounds is concerned. Toxicological investigations are carried out mainly in the laboratories of industry.

To conclude what has been said about preclinical development, it must be mentioned that a sufficient protection by patent is also an important prerequisite for the development of a new drug. Almost invariably this can only be done in cooperation with a competent pharmaceutical company. Such companies do not have any financial interest in substances which are not sufficiently protected by patent (Fig. 7).

If all preclinical experiments have shown satisfactory results and registration with the national drug authorities has been completed, clinical development may begin. This is subdivided into three phases (Fig. 7). Phase I is essentially a toxicology study in humans where the side-effects of the new drug are examined in tumor patients who have a low life expectancy of only about half a year. In phase II antitumor activity is actually screened. Phase III is designed to determine the ultimate therapeutic value of a new drug. On the basis of these methods the development of new tumor-inhibiting metal complexes will be reported on in the following.

SYNTHESIS OF DIRECT DERIVATIVES OF CISPLATIN

BASIC IDEAS

The relation between chemical structure and antitumor activity of platinum compounds has been studied extensively. In part it had already been created by Barnett Rosenberg. Tumor-inhibiting platinum compounds have a precisely defined structure:

Chemical structure of tumor-inhibiting platinum compounds:

$$cis\text{-}A_2Pt^{II}X_2$$
or
$$cis\text{-}A_2Pt^{IV}X_2Y_2$$

A = a cis-orientated organic amine ligand, preferably a primary amine, or ammonia
A_2 = a bidentate amine
X = an easily replaceable group, such as Cl^- or carboxylate
Y = two additional trans-orientated groups such as OH^- or Cl^-

Corresponding trans-configurated compounds are inactive.
Active complexes are usually neutral.

Meanwhile several thousand cisplatin derivatives have been synthesized, and, of these, more than thousand have been tested preclinically. About ten of these have qualified for clinical studies in patients. Much to the disappointment of clinicians, however, it turned out that the spectrum of activity of these new complexes was relatively similar to that of cisplatin, except for minor improvements. Some success was achieved in decreasing nephrotoxicity, particularly with carboplatin and iproplatin (Fig. 10). Carboplatin also causes less nausea. However, while showing roughly the same therapeutic efficacy, these compounds have myelotoxicity as a dose-limiting factor, and this to an even greater extent than cisplatin. This myelotoxicity is sometimes hard to control, because it may occur even after the disruption of therapy. Drug authorities in many countries have meanwhile permitted carboplatin for routine clinical use (Curt et al. 1986; Krakoff, 1988; Achterrath et al. 1984; Vermorken et al. 1984, 1985; Sternberg et al. 1984; Rose and Schurig, 1985; Harrap, 1985).

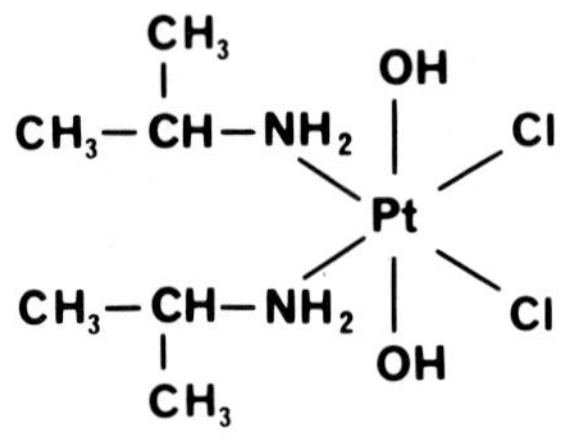

cis-Dichloro-trans-dihydroxydi(isopropylamine)platinum(IV)
Iproplatin

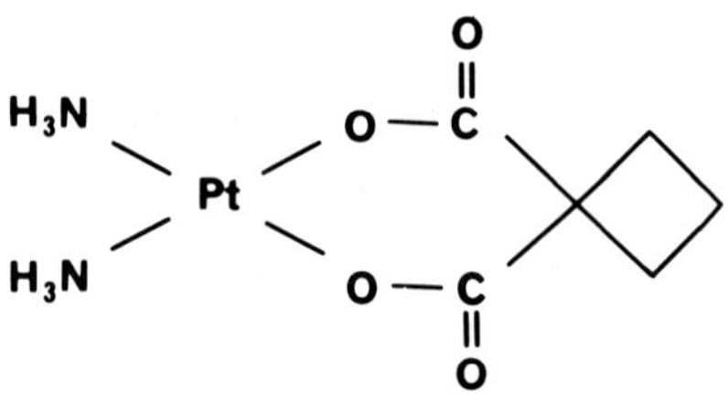

cis-Diammin(cyclobutane-1,1-dicarboxylato)platinum(II)
Carboplatin

Fig. 10. Iproplatin and carboplatin.

Cisplatin could never really be sufficiently protected by patent, particularly in Europe. This is, however, possible in the case of the new complexes. The development of substances which cannot be sufficiently protected by patent is not profitable, and certainly not for large companies. This may explain some of the activity in the sector of tumor-inhibiting cisplatin derivatives. Despite all this, experience seems to confirm the fact that the synthesis of direct derivatives of cisplatin only leads to side-effects being reduced. The spectrum of indication can hardly be expanded.

THE DEVELOPMENT OF NON-PLATINUM COMPLEXES

The synthesis of non-platinum complexes is likely to yield much fewer tumor-inhibiting substances than that of direct cisplatin derivatives. After all, precise structure-activity relations are known for the platinum derivatives, with the result that minor variations in synthesis will easily lead to other active complexes. On the other hand, the successful synthesis of non-platinum complexes, i.e., when a tumor-inhibiting compound has been found, will more frequently lead to compounds that have a completely different spectrum of indication than platinum compounds have, owing to the inevitable change in chemical properties. Substitution of the central metal hence presents more opportunities for obtaining complexes which are active against tumors that can hardly be treated and that account for the major share of cancer mortality today.
Thus several hundred complexes have been synthesized. In order to assess their future clinical potential, they are best subdivided into drugs which are at a preclinical stage of development and drugs which have already qualified for clinical studies.

NON-PLATINUM COMPLEXES IN PRECLINICAL TRIALS

Today a large number of complexes are known that show antitumor activity in certain preclinical models. Of course the search for new compounds at first centered on the immediate surroundings of platinum, i.e., among the platinum metals. These include, besides platinum itself, also ruthenium, rhodium, palladium, osmium, and iridium. Ruthenium compounds, above all, have gained considerable importance here. Other developments in the field of tumor-inhibiting non-platinum complexes which are still in preclinical trials and have not yet qualified for clinical studies will be described in the following. The tumor-inhibiting palladium complexes will be mentioned first.

Fig. 11 shows 1,2-Diaminoethanedinitratopalladium(II), the basic structure of some tumor-inhibiting palladium complexes. At a dose of 80 mg/kg, this complex produced almost a doubling of survival time of treated animals (T/C = 194 %) bearing the sarcoma 180 ascitic tumor. The compound, however, was applied intraperitoneally in a single dose on the same day the tumor was transplanted (also i.p.). When applied intraperitoneally the tumor should normally be allowed to grow over 24 hours at least before therapy is begun (Gill, 1984).

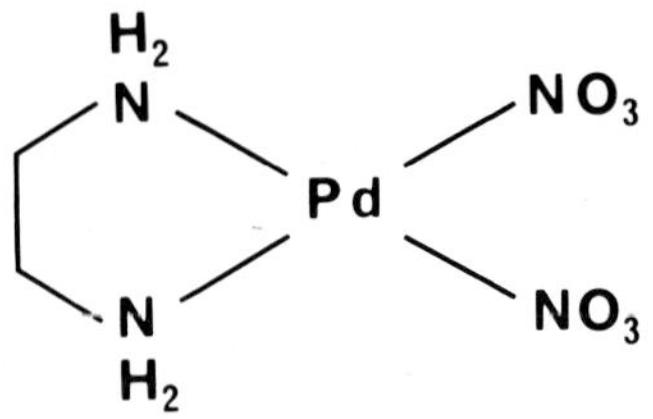

1,2-Diaminoethanedinitratopalladium(II)

Fig. 11. 1,2-Diaminoethanedinitratopalladium(II).

Other, similar configurated compounds, which have as ligands, among others, diaminocyclohexane, 1,2-diaminopropane, 1,3-diaminopropane, 2,2'-bipyridine, or ammonia, and as hydrolizable groups malonate, oxalate, glutarate, and cyclobutanedicarboxylate, instead of NO_3^-, were less active under the same conditions and sometimes even inactive. Hence tumor-inhibiting palladium complexes have as yet not shown any substantial advantages over cisplatin.

Other tumor-inhibiting structures from the field of platinum metals include the dirhodiumtetracarboxylates (Fig. 12). Rhenium, which does not belong to the platinum metals, yields isostructural complexes which can be seen in the same Figure. The rhodium compounds showed statistically significant activity in the Ehrlich ascitic tumor, the sarcoma 180 ascitic tumor, and in the intraperitoneally transplanted leukemias P 388 and L 1210 (Erck et al. 1974). If acetate, proprionate, and butyrate are used as carboxylates, an increase in antitumor activity can be observed in the following order: acetate < proprionate < butyrate. Antitumor activity hence increases with the lipophilia of the compounds. This effect can frequently be observed with oncostatic metal complexes (Howard et al. 1977). The rhenium complex showed significant activity in the intraperitoneally transplanted leukemia P 388, in the subcutaneously transplanted sarcoma 180, and in the s.c. transplanted melanoma B 16. Bis-µ-proprionatodiaquatetrabromodirhenium(II), the complex shown in the Figure, proved to be superior to similarly structured complexes with acetate and butyrate ligands and chloride as halogen ligand (Dimitrov and Eastland, 1978; Eastland et al. 1983).

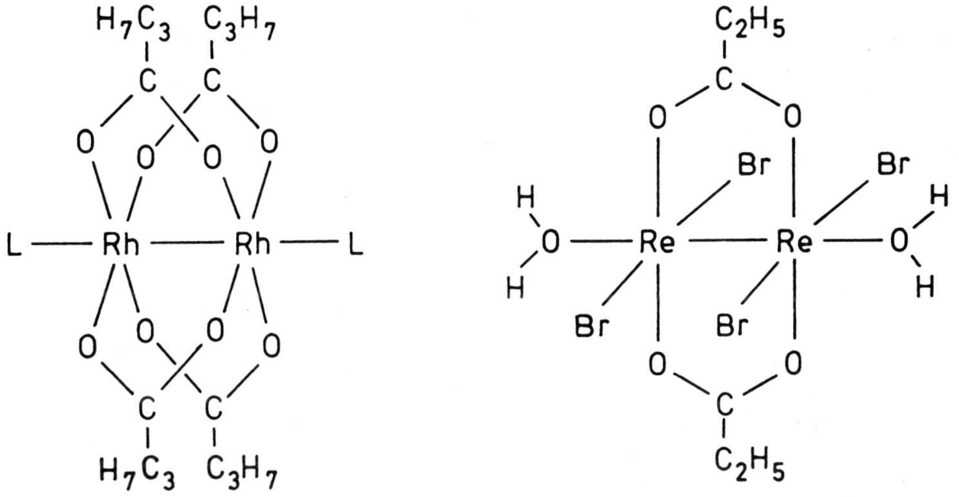

Fig. 12. Tetra-µ-butyratodirhodium(II) and Bis-µ-propionatodiaquatetrabromorhenium(II).

The cyclooctadien complexes of rhodium and iridium (Fig. 13) have certain antitumor activity in the subcutaneously transplanted Lewis lung carcinoma, with the iridium compound not showing any inhibiting effect on metastases, in contrast to the rhodium complex. Both compounds have a clear effect also on the Ehrlich ascitic tumor (Sava et al. 1983).

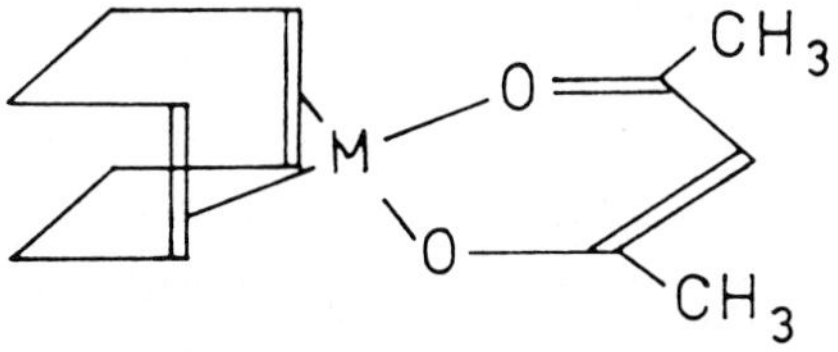

Pentane-2,4-dionatocycloocta-1,5-dien complexes of rhodium(I) and iridium(I)

Fig. 13. Pentane-2,4-dionatocycloocta-1,5-dien complexes of rhodium(I) and iridium(I).

Interest in tumor-inhibiting gold complexes (Fig. 14) was aroused by the cytotoxic activity of auranofin in vitro. Auranofin is well-known as a drug for primary chronic polyarthritis (Fig. 2). However, its promising activity in cell cultures could not be reproduced in vivo. Antitumor activity was observed, however, with other gold complexes, particularly those with 1,2-Bis(diphenylphosphino)ethane (dppe) as ligand. This ligand also shows antitumor activity, which is reinforced by complexation with gold. ClAu(dppe)AuCl reaches T/C values of about 200 % in the P 388 leukemia. In addition, significant activity was observed in the L 1210 leukemia, the M5076 reticulum cell carcinoma, the B 16 melanoma, the mammary adenocarcinoma 16/c (all of these are i.p. transplanted and i.p. treated), and the ADJ/PC6 blastocytoma (subcutaneously transplanted). $Au(dppe)_2Cl$ is also active, showing comparable activity in the P 388 leukemia, but it is no longer active when given intravenously, subcutaneously, or p.o., i.e., activity can only be demonstrated with local application to the tumor (ip/ip model). There was also no activity when the P 388 leukemia was transplanted intravenously and when the complex was given i.p. or i.v. afterwards. T/C values of about 160 % were found in the i.p. transplanted M 5076 reticulum cell carcinoma, and T/C values of about 140 % were reached in the B 16 melanoma. Also, activity could be observed in the subcutaneously transplanted mammary adenocarcinoma 16/c. Amazingly it was found that a gold complex, potassium[dicyanogold(I)], had good activity against the Lewis lung carcinoma, reaching T/C values of 200 % and producing a high percentage of cured animals when it is applied i.p. from day 1 to 9 at a dose of 2.5 mg/kg. However, comparatively promising activity could not be reproduced in other tumor models. As far as sensitivity of certain tumor models to specific metal complexes is concerned, which have little predictivity for later clinical activity when taken on their own, this is a case in point. At the beginning of the 20th century, the compound was tested in the treatment of tuberculosis, but it was soon discarded as a drug for this indication on account of its high toxicity (Simon et al. 1981; Berners-Price et al. 1986; Sadler et al. 1984; Berners-Price and Sadler, 1988).

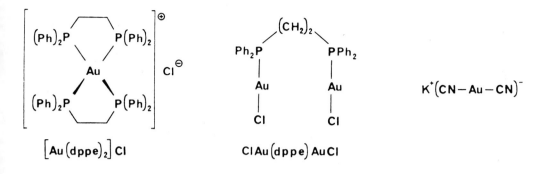

Fig. 14. Bis[1,2-bis(diphenylphosphino)ethane]gold(I)chloride, 1,2-Bis(diphenylphosphino)ethanebis[gold(I)chloride], and Potassium[dicyanogold(I)].

In synthesis of tumor-inhibiting copper complexes, 1,2-Bis(diphenylphosphino)ethane was also used as ligand. However, much more promising compounds were obtained with the macrocyclic ligand tetrabenzotetrazacyclohexandecine (Fig. 15). This complex reached T/C values of 170 % in the melanoma B 16 and of 140 % in the leukemia P 388. Apart from the copper complexes with thiosemicarbazones, the bis(2-hydroxybenzaldoximato)copper(II) complexes are also supposed to have certain antitumor activity. The latter were active in the i.p. treated Ehrlich ascitic tumor, but not in the P 388 leukemia in vivo (Fig. 15). Further data on the chemical structure of these compounds would be interesting (Lumme et al. 1984; Sadler et al. 1984; Elo and Lumme, 1985; Lumme and Elo, 1985).

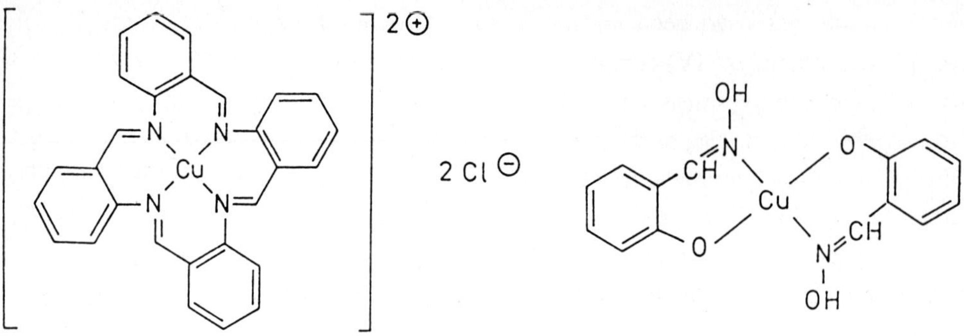

Fig. 15. [Tetrabenzo(b,f,j,n)(1,5,9,13)tetrazacyclohexadecanecopper(II)]dichloride and Bis(1,2-hydroxybenzaldoximato)copper(II) complexes.

2-Hydroxybenzalanilinato complexes of cobalt are known, which are able to reduce growth of the intramuscularly transplanted Walker 256 carcinosarcoma (Fig. 16) (Hodnett et al. 1971).

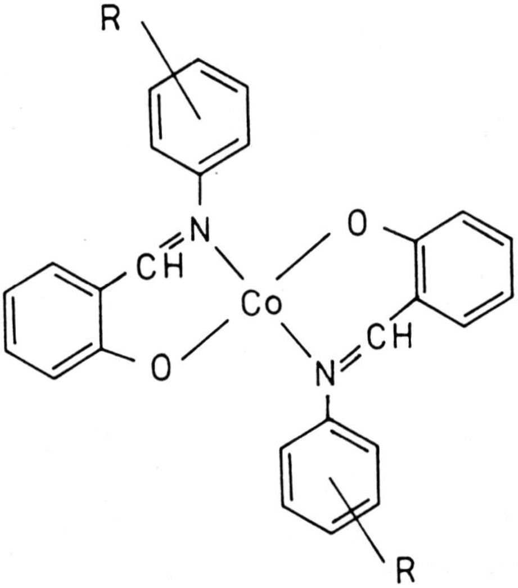

Fig. 16. Bis(2-hydroxybenzanilinato)cobalt(II) complexes.

Numerous studies have been made into the antitumor activity of Dihalogenobis(η^5-cyclopentadienyl)metal(IV) complexes of the general structure Cp_2MX_2 (Fig. 17), and also of metallicenium salts such as $(Cp_2M)^+X^-$ with X = acidoligand, and of uncharged metallocenes, Cp_2M, e.g., of the main group elements tin and germanium. It is probably Dichlorobis(η^5-cyclopentadienyl)titanium(IV) which plays the most important role. The antitumor acitivity of this complex has been studied extensively in the Ehrlich ascitic tumor, the melanoma B 16, the colon 38 carcinoma, and the Lewis lung carcinoma, as well as in numerous human tumor xenografts (Köpf-Maier and Köpf, 1988).

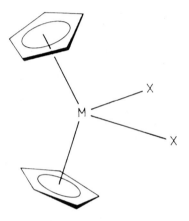

Fig. 17. Dihalogenobis(η^5-cyclopentadienyl)metal(IV) complexes.

Tin compounds will be reviewed extensively in other contributions to this congress and will thus not be discussed here. Nevertheless it must be mentioned that they play an important role in the preclinical screening of new antitumor agents.

On the whole, tin compounds, metallocenes, tumor-inhibiting gold compounds, and ruthenium complexes have been studied most extensively. The other types of substances mentioned are still at a fairly early stage of preclinical development. The goldphosphine complexes and metallocenes mentioned above are active in a broader spectrum of experimental tumors. The latter showed an interesting activity against colonadeno-xenografts in nude mice, which can be supposed to have certain predictivity for the clinic. Among the non-platinum complexes which are under preclinical examinations, the ruthenium compounds are fairly advanced. They will be discussed in the following.

TUMOR-INHIBITING RUTHENIUM COMPLEXES

Although the tumor-inhibiting activity of some classic ruthenium complexes such as "ruthenium red", cis-Ru(DMSO)$_4$Cl$_2$, or cis-[Ru(NH$_3$)$_4$Cl$_2$]Cl, (Fig. 18), has been known for a relatively long time now, none of these complexes and none of their derivatives have so far been able to qualify for clinical trials (Giraldi et al. 1974, 1977; Clarke, 1980; Tsuruo et al. 1980; Clarke et al. 1988; Alessio et al. 1988; Sava et al. 1983, 1984, 1985). The reasons for this may be that in biological experiments the compounds were hardly ever investigated in realistic and sophisticated tumor models. So far ruthenium complexes have almost always been tested in the P 388 leukemia and compared to cisplatin. Since this tumor model is very sensitive to platinum compounds, it would thus be surprising to find ruthenium complexes showing better effects than cisplatin does.

It must be taken into account that the P 388 model - just as other primary screening methods - is suited for separating active complexes from inactive ones rather than for weighing up the statuses of different groups of substances such as platinum complexes and ruthenium complexes. Hence this is by no means an exclusive criterion, as is often supposed when tumor-inhibiting ruthenium complexes do not show a higher activity in this model than cisplatin does. They may just as well qualify for further development.

Within a class of compounds such as that of ruthenium complexes, however, a transplantable tumor such as the P 388 leukemia is fairly well suited for distinguishing activity grades, because all the single derivatives from a class of substances probably act on the basis of a similar molecular mechanism. If then a highly active structure type is found among the many ruthenium compounds, it will be necessary to resort to sophisticated and clinically predictive models in order to be able to assess the potential advantages of therapy with a particular derivative.

One of the two best-studied ruthenium derivatives is cis-dichlorotetrakis(dimethyl-sulfoxide)ruthenium(II), cis-Ru(DMSO)$_4$Cl$_2$ (Fig. 18). This complex is fairly soluble in water. It shows only marginal activity against the P 388 leukemia, but it has good metastases-inhibiting effects on the Lewis lung tumor. Still less active is ruthenium red. Cis-Tetrammindichlororuthenium(III)chloride (Fig. 18), cis-[Ru(NH$_3$)$_4$Cl$_2$]Cl, is active against the P 388 leukemia and reaches T/C values of about 160 % in this model. It is also fairly well soluble in water (Clarke, 1980).

A third derivative, fac-trisamminetrichlororuthenium(III), fac-Ru(NH$_3$)$_3$Cl$_3$ (Fig. 18), is highly active in the P 388 leukemia, compared to the tetrammine complex, but in contrast to the latter, it is insoluble in water (Clarke, 1980).

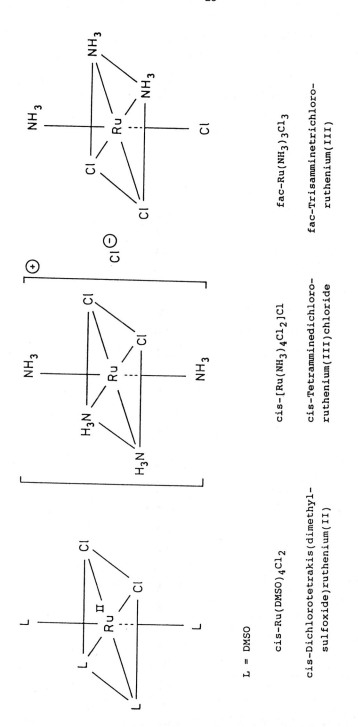

Fig. 18. Three ruthenium complexes with known antitumor activity.

In contrast to many other structure classes of tumor-inhibiting metal complexes, it is thus possible to synthesize active and water-soluble derivatives in the field of ruthenium complexes. This should be kept in mind when synthesizing new complexes. A certain water-solubility is very advantageous for later application of these compounds in the clinic. The adding of solubilizers and the carrying out of sophisticated galenic procedures with complexes that are insoluble in water usually causes additional problems as to the analytics of the galenic formulation and hence for registration with the national drug authorities. Decomposition of the metal complexes can also be accelerated by adding solubilizers, or it may produce adducts with the active complex which are difficult to characterize.

Proceeding from the ruthenium compounds described above, some promising new representatives have been synthesized in Heidelberg during the last few years. These will be described in the following. In order to be able to assess the activity of these ruthenium complexes, above all in comparison to those known from the literature, we have tested some of the typical representatives in the P 388 model (Keppler and Rupp, 1986; Keppler et al. 1987, 1989; Garzon et al. 1987; Berger et al. 1989). The results of this prescreening are summarized in Table 1. They convey an idea of the possible variations in the activity spectrum.

	Water-solubility	T/C (%)
$(HB)_3Ru^{III}Cl_6$	+	100-120
$(HB)_2(Ru^{IV}Cl_6)$	+	100-120
$(HB)_4(Ru^{IV}_2OCl_{10})$	+	100-120
$Ru^{II}(DMSO)_4Cl_2$	+	125
$Ru^{II}(DMSO)_{4-n}B_nCl_2$	+	100-125
$Ru^{III}(RSR)_3Cl_3$	−	130-135
$Ru^{III}B_3Cl_3$	−	100-130
$Ru^{II/III}_2(O_2CR)_4Cl$	+	125-135
cis-$[Ru^{III}(NH_3)_4Cl_2]Cl$	+	160*
fac-$Ru(NH_3)_3Cl_3$	−	190*
HB-trans-$[Ru^{III}B_2Cl_4]$	+	140-200
$(HB)_2[Ru^{III}BCl_5]$	+	140-200

Table 1: Comparison of the antitumor activity of different classes of ruthenium compounds in the P 388 leukemia; B = nitrogen heterocycle, R = organic ligand; n = 1-4; * in accordance with Clarke, 1980.

The first three complexes in Table 1 are rutheniumhexachlorides in the oxidation stages III and IV and oxygen-bridged ruthenium chlorides with different, protonated heterocycles as cations. They are inactive, with T/C values < 125 % (Keppler et al. 1989).

The synthesis of derivatives of the ruthenium complex Ru(DMSO)$_4$Cl$_2$ (Table 1) with the aim of obtaining compounds of the type Ru(DMSO)$_{4-n}$B$_n$Cl$_2$ (B = heterocycle) was successful only in the case of pyridine and pyridazine derivatives. The resulting compounds had little tumor-inhibiting activity, with T/C values below 125 %. Hence they do not show any advantages over Ru(DMSO)$_4$Cl$_2$ itself (Keppler et al. 1989).

The next two complexes in Table 1 are derivatives of fac-Ru(NH$_3$)$_3$Cl$_3$. Like the latter, they are insoluble in water. Two complexes of the general formula Ru(RSR)$_3$Cl$_3$ - with RSR = phenylmethylthioether and the corresponding derivative with ethyl instead of phenyl - could be obtained. Both complexes are active in the P 388 tumor system and in the sarcoma 180. However, the T/C values of 130 % obtained in the P 388 tumor are not promising. Much more promising results were obtained in the sarcoma 180 ascitic tumor model, with T/C values up to 300 %. This model, however, seems to be comparatively sensitive to tumor-inhibiting ruthenium complexes. Many of the tested substances reach T/C values of > 300 %. This sounds encouraging and promising at first sight, but one has to be aware of the fact that it is not the aim of such experiments to cure tumor-bearing mice but to find antitumor activity grades within a number of different complexes. This cannot be done if all substances turn out to be highly active owing to an inherent sensitivity of a particular tumor system. More information will be obtained if an experimental model is chosen that will yield medium activity for the majority of compounds. Then it will be possible to distinguish species which exceed this medium activity and compounds which have distinct below-average activity (Keppler et al. 1989). Particularly in the case of ruthenium complexes, the P 388 leukemia is a "critical" model in this sense. Selection of the tumor model to be favored thus highly depends on the central metal or the class of compounds to be examined. The P 388 leukemia, for example, is virtually insensitive to titanium compounds, with the result that the sarcoma 180, with its "medium" sensitivity to this type of complexes, was chosen as critical test.

The activity loss of the thioether complexes as against the trisammine complex may be attributed to the different molecular structure. Ru(NH$_3$)$_3$Cl$_3$ is facially configurated whereas the thioether complexes are meridionally configurated. This could be proved by x-ray analyses (Keppler et al., 1989).

Antitumor activity of the compounds of the general formula RuB$_3$Cl$_3$ (B = heterocycle) (see Table 1) is not very promising. Only the derivatives with 4-dimethylaminopyridine, pyridine-4-aldehyde, and pyridazine reached T/C values of 125 %, which indicate biological activity. All T/C values obtained are clearly below that of the trisammmine complex, which served as standard. This may be attributed to the fact that these derivatives, just as the thioether complexes, are not facially but meridionally configurated, as we were able to prove (Keppler et al. 1989).

We have then concentrated our attention on water-soluble ruthenium complexes and have studied such complexes as have not yet been considered under the aspect of antitumor activity. These include, among others, binuclear carboxylates of ruthenium, which were selected because of the well-known tumor-inhibiting activity of the binuclear carboxylates of rhodium and rhenium. The schematic structures and the T/C values of two ruthenium complexes in the P 388 system are shown in Fig. 19. Acetate and proprionate served as ligands. Both substances are water-soluble and faintly but significantly active, with T/C values of 125 and 133 % (Keppler et al. 1989).

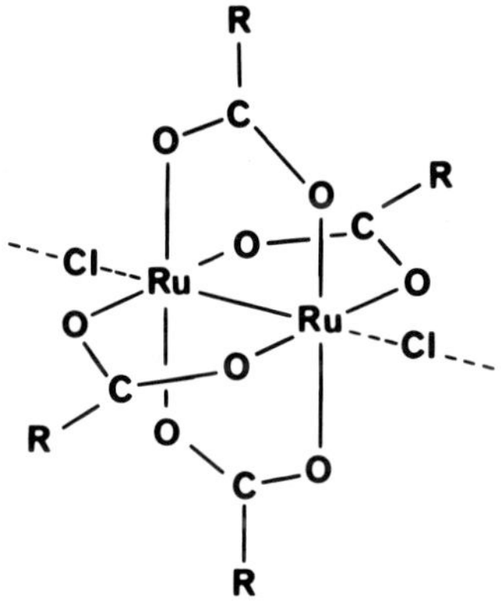

R = CH$_3$ or C$_2$H$_5$

	Dose mmol/kg	Dose mg/kg	Therapy on days	T/C (%)
Ru$_2$(OOCCH$_3$)$_4$Cl	0.21	100	1	125
Ru$_2$(OOCC$_2$H$_5$)$_4$Cl	0.16	85	1	125
	0.06	32	1,5,9	133

Fig. 19. Antitumor activity of two dirutheniumtetracarboxylates against the P 388 leukemia; T/C (%) = (median survival time of treated animals / median survival time of control animals) x 100.

The water-soluble ruthenium species of the general formulas HB(RuB$_2$Cl$_4$) and (HB)$_2$(RuBCl$_5$) (Fig. 20), which have been developed by us, show much better activity. Some of the representatives of these classes even reach T/C values of 140-200 % in the P 388 leukemia. The heterocycles in the anions are mostly trans- orientated. Derivatives with only one heterocycle in the anion are usually a little less active than their corresponding counterparts with two heterocycles. Synthesis of these compounds is carried out with purified ruthenium(III)chloride and the corresponding heterocycles in a solution of hydrochloric acid or of ethanole/hydrogen chloride. The reaction is fairly difficult to produce and depends on variables such as pH, temperature, and concentration (Keppler et al. 1989).

Complexes of the type HB(RuB$_2$Cl$_4$) can in principle be cis- or trans- configurated. The trans- configuration was demonstrated by means of various methods of spectroscopy such as NMR, IR, Mössbauer spectroscopy, as well as by x-ray analysis (Keppler et al. 1987, 1989).

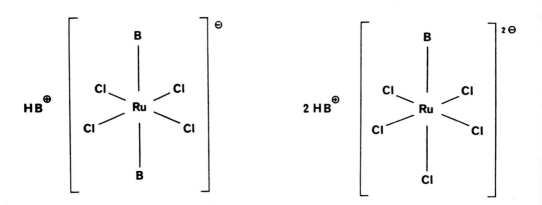

Fig. 20. General structure of the tumor-inhibiting ruthenium compounds of the general formulas HB(RuB$_2$Cl$_4$) and (HB)$_2$(RuBCl$_5$); B = heterocycle.

Table 2 illustrates the results obtained with a number of these complexes in the P 388 leukemia in comparison to the clinically established drugs cisplatin and 5-fluorouracil. HIm(RuIm$_2$Cl$_4$), ICR, trans-imidazolium-bisimidazoletetrachlororuthenate(III), reached T/C values of about 200 %, compared to T/C values of about 180 % in the case of cisplatin, and about 150 % in the case of 5-fluorouracil. The T/C values of the other substances range between 130 % for the aminothiazole compound and 160 % for the chinoline compound. Comparison of the various therapy schemes revealed certain advantages of treatment on the days 1, 5, and 9 after tumor transplantation (Keppler et al. 1989).

Table 3 compares the results of ICR in different transplantable tumor models. The column in the middle represents the evaluation parameter - survival time or tumor weight. T/C values must be interpreted as explained above. Apart from showing good activity in the P 388 leukemia, ICR reaches T/C values of about 250 % in the Walker 256 carcinosarcoma, and of about 150 % in the Stockholm ascitic tumor. Survival times of > 300 % could be achieved in the MAC 15A tumor, a transplantable colon adenocarcinoma (Keppler et al. 1989; Berger et al. 1989).

The subcutaneously transplanted melanoma B 16 and the intramuscularly transplanted sarcoma 180 - both are solid tumors - were treated intravenously in order to provide proof of systemic activity. Therapy resulted in a significant reduction of tumor volume to 15 % and 45 %, respectively, compared to the control animals (100 %).

	Dose	Treatment on days	T/C (%)	range
controls	-	-	100	(100-137)
Cisplatin	3 mg/kg 0.01 mmol/kg	1,5,9	175	(100-275)
5-FU	60 mg/kg 0.46 mmol/kg	1,5,9	144	(115-179)
ICR, $ImH(RuIm_2Cl_4)^*$	209.3 mg/kg 0.45 mmol/kg	1	156	(138-200)
	69.8 mg/kg 0.15 mmol/kg	1,5,9	194	(138-262)
	23.3. mg/kg 0.05 mmol/kg	1-9	163	(138-225)
$(ImH)_2(RuImCl_5)^*$	218.5 mg/kg 0.45 mmol/kg	1	150	(100-150)
	72.8 mg/kg 0.15 mmol/kg	1,5,9	163	(150-163)

continued on next page

(ImH)$_2$(RuImCl$_5$)*	24.3 mg/kg 0.05 mmol/kg	1-9	156	(150-163)
(1MeImH)$_2$-[Ru(1MeIm)Cl$_5$]	52.4 mg/kg 0.1 mmol/kg	1,5,9	144	(78-144)
(4MeImH)-[Ru(4MeIm)$_2$Cl$_4$]*	73.5 mg/kg 0.15 mmol/kg	1,5,9	133	(122-155)
(BzImH)[Ru(BzIm)$_2$Cl$_4$]*	60.7 mg/kg 0.1 mmol/kg	1,2,3	155	(144-178)
(BzImH)$_2$(RuBzImCl$_5$)	100.7 mg/kg 0.15 mmol/kg	1,5,9	133	(122-144)
(PzH)[RuPz$_2$Cl$_4$]	44.8 mg/kg 0.1 mmol/kg	1,2,3	144	(144-156)
(DiMePzH)-[Ru(DiMePz)$_2$Cl$_4$]	53.2 mg/kg 0.1 mmol/kg	1,2,3	133	(111-189)
(IndH)[RuInd$_2$Cl$_4$]*	91.1 mg/kg 0.15 mmol/kg	1,5,9	133	(133-144)
TrH(RuTr$_2$Cl$_4$)	73.1 mg/kg 0.15 mmol/kg	1,5,9	138	(115-150)
(ChinH)[RuChin$_2$Cl$_4$]	64.0 mg/kg 0.1 mmol/kg	1,2,3	160	(130-270)
(2AmiThiazolH)-[Ru(2AmiThiazol)$_2$Cl$_4$]*	54.4 mg/kg 0.1 mmol/kg	1,2,3	130	(130-130)

Table 2: Antitumor activity of selected ruthenium complexes of the general formulas HB(RuB$_2$Cl$_4$) and (HB)$_2$(RuBCl$_5$) in the P 388 leukemia, compared to cisplatin and 5-fluorouracil. All T/C values are statistically significant, in contrast to that of controls (Steel Test). * These complexes were also made available to the NCI, where activity in the P 388 leukemia was confirmed.

Tumor model	Evaluation parameter	optimum T/C value (%)
P 388 leukemia	ST	200
Walker 256 carcinosarcoma	ST	250
Stockholm ascitic tumor	ST	150
B 16 melanoma, s.c. growing	TW	15
sarcoma 180 i.m. growing	TW	45
MAC 15A tumor	ST	>300
AMMN-induced colorectal tumors of the rat	TW	10

Table 3: Survey of the antitumor activity of ICR in different experimental, both transplantable and autochthonous, tumor systems; ST = median survival time, TW = median tumor weight, AMMN = acetoxymethylmethylnitrosamine.

As can be seen from the general organization scheme (Fig. 7), the next important step in preclinical trials, following these encouraging results, must include considerably more sophisticated models in order to delimit the clinical spectrum of indication of the new substances. It was in the course of these investigations that the good activity of ICR in an autochthonous tumor model, the AMMN-induced colorectal tumors of the rat, was found. It has been illustrated that this is a particularly valuable model, which is well suited for predicting the clinical activity of new compounds in this type of tumor. The tumors are induced by the carcinogen acetoxymethylmethylnitrosamine (AMMN), and they are macroscopically and microscopically very similar to the human tumors. Fig. 21 shows these tumors in the colon of a rat, together with the structure of the carcinogen that induces these tumors. The sensitivity of these tumors to chemotherapeutic agents is almost the same as that of the corresponding human tumors. The tumors are not sensitive to cisplatin therapy and do not respond to treatment with alkylating agents such as cyclophosphamide. At present 5-fluorouracil is the only drug in clinical use to produce a certain reduction of tumor volume. This effect can also be reproduced in the experimental model (Keppler et al. 1989; Berger et al. 1989).

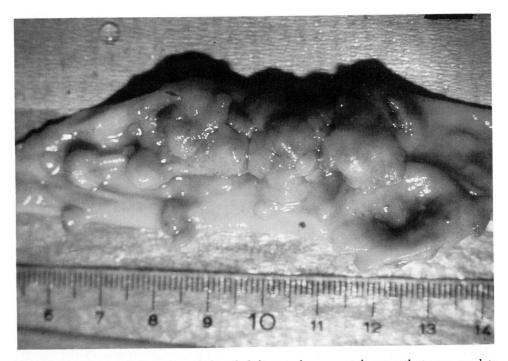

Fig. 21. Structure of acetoxymethylmethylnitrosamine, a carcinogen that was used to induce the colon tumors shown below. These tumors developed in the colon of a rat 20 weeks after the carcinogen was applied for the first time. The multiple adenotumors resemble human tumors already from their outward appearance.

Apart from ICR, some other ruthenium derivatives have been tested in this model. Besides ICR, trans-indazolium-bisindazoletetrachlororuthenate(III), HInd(RuInd$_2$Cl$_4$), shows the best activity. The results obtained with these two compounds in comparison to cisplatin and 5-fluorouracil are given in Fig. 22. Three different experiments are summarized there. The tumor volume of the control animals was always standardized to 100 %, with the result that the therapeutic efficacy of the different substances can directly be compared. Cisplatin is completely inactive in this model, just as it is in the clinical tumor type. The well-known positive effect of 5-fluorouracil on comparable clinical tumors could also be reproduced in this model. Tumor volume decreased to 40 %, compared to that of controls. ICR, given at a dose of 7 mg/kg twice a week over ten weeks, reduced tumor volume to 20 % in one case and to 10 % in the other. The indazole derivative - HInd(RuInd$_2$Cl$_4$) - had already turned out to be less toxic in chronic application, and hence it could be applied at a higher dose. The result was a decrease in tumor volume to even 5 %. This equals a reduction of tumor mass of 95 %. Final evaluation showed that in this group about one third of the animals were even tumor-free.

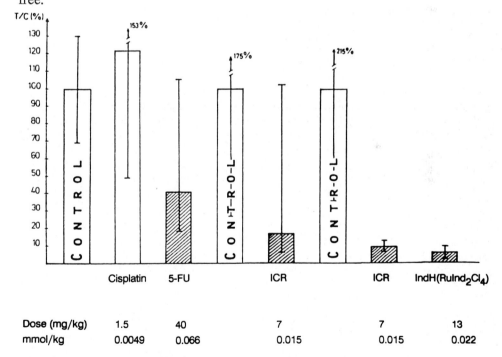

Fig. 22. Test results of the two ruthenium compounds ICR = HIm(RuIm$_2$Cl$_4$) and HInd(RuInd$_2$Cl$_4$) in autochthonous colorectal tumors of the rat. Three different experiments are summarized here. The tumor volume of each control group is indicated. It has been standardized to 100 %. Doses were applied twice a week over ten weeks. The reduction of tumor volume represented by the shaded columns is statistically significant compared to the control groups.

Fig. 23 gives additional information on the last experiment. Apart from tumor volume, the parameters of weightloss and mortality are taken into account here. Mortality and weightloss are caused by invasive tumor growth. The application of ICR resulted in a 90 % reduction of tumor weight, while mortality remained unchanged in comparison to controls. However, weightloss was more pronounced. This therapeutic effect, along with still tolerable toxicity, could not be improved by applying higher doses. The application of 10 mg/kg of ICR - twice a week over ten weeks - was associated with a weightloss of 40 % and a mortality rate of 55 %. The drastic effect of the indazole derivative - $HInd(RuInd_2Cl_4)$ - was manifested in a 95 % reduction of tumor volume, no mortality (0 %), and virtually no weightloss in comparison to controls (- 6 %). This is evidence of a successful and non-toxic therapy of colorectal tumors with this drug. Detailed evaluation showed that one third of the animals were completely tumor-free, and thus cured. These findings are extremely remarkable insofar as cisplatin is completely inactive in this type of tumor, and because these colorectal tumors account for a high percentage of human cancer mortality. Thus these preclinical results give rise to the hope of promising future activity in the clinic (Garzon et al. 1987; Keppler et al. 1989; Berger et al. 1989).

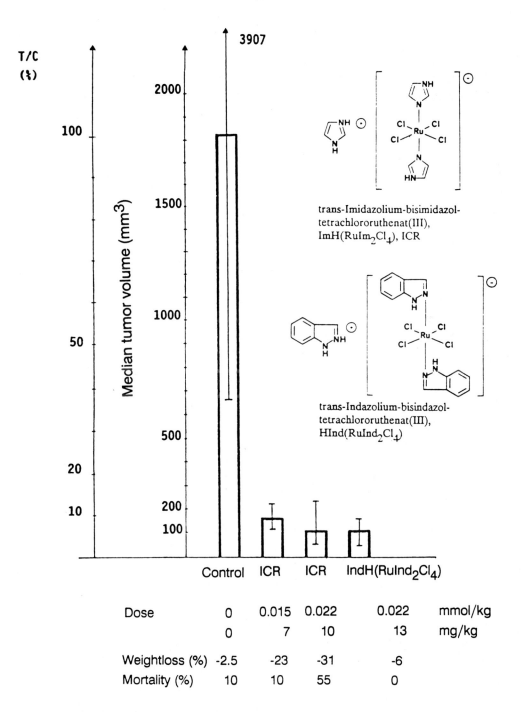

Fig. 23. Comparison of the activity of the two ruthenium compounds ICR = HIm(RuIm$_2$Cl$_4$) and HInd(RuInd$_2$Cl$_4$) in autochthonous colorectal tumors of the rat. Therapy was carried out twice a week over ten weeks.

Stability of the galenic formulation of ICR and its derivatives is an important aspect particularly in connection with clinical use. HPLC investigations showed that the half-life of the compound is roughly 400 minutes. This means that during the first thirty minutes more than 95 % of the complex remains undecomposed. This stability is quite sufficient for infusion therapy in the clinic.

Finally we would like to describe toxicological results with ICR, HIm(RuIm$_2$Cl$_4$), and HInd(RuInd$_2$Cl$_4$). The LD$_{50}$ of ICR in single intravenous application is about 125 mg/kg in mice when the application volume is 2 ml of physiological saline per 100 g of mouse. In later experiments we found a high dependence of mortality on the volume of solubilizer applied. A dose of 100 mg/kg of ICR is associated with 40 % mortality if the drug is applied in 2 ml of solubilizer per 100 g of mouse. Mortality, however, can be reduced to 0 % when the volume of solubilizer is increased to 8 ml (= 2 ml in a mouse of 25 g). Mortality is 100 % when 200 mg/kg of ICR are applied, but this mortality can be reduced to 25 % by the same dilution procedure. Similar effects can be observed when the indazole compound is used. The LD$_{50}$ increases from 50 mg/kg to 100 mg/kg when the same dilution is prepared. According to this data, nephrotoxicity, which is common to both compounds, can be reduced significantly by applying higher volumes of liquid.

Ultrastructural investigations have shown that the liver and the kidney are the main target organs of drug toxicity at a dose of 110 mg/kg of ICR and 2 ml of solubilizer per 100 g of mouse. Hematological findings include erythropenia (decrease in the number of red blood cells) and an increase in creatinine and liver enzymes.

In examining chronic toxicity both the imidazole and the indazole compound were applied twice a week at equimolar doses. ICR was administered at 10 mg/kg twice a week over seven weeks, and the indazole compound at 13 mg/kg twice a week over eight weeks. There was no mortality in either experiments. However, the animals treated with ICR suffered from a significant weightloss in the last two weeks, which was caused by drug toxicity and which led to therapy being disrupted after the seventh week. In contrast to this, no toxicity could be observed with the indazole compound up to the eighth week. Thus the indazole compound is better tolerated in chronic application than the imidazole compound. These results were not to be expected after the studies into acute toxicity. In this case, the situation is reversed. The advantages of long-term application of the indazole compound have also proven useful in the treatment of colorectal tumors (Keppler et al. 1989; Berger et al. 1989).

NON-PLATINUM COMPLEXES IN CLINICAL STUDIES

Drugs which are already in clinical studies or in clinical use are considerably more advanced in terms of development. The high standards required of a new drug to be admitted to clinical studies may be recognized from the fact that in the last few decades only four non-platinum complexes have been tested clinically on an international level. These include germanium-132, carboxyethylgermaniumsesquioxide; spirogermanium, N-(3-dimethylaminopropyl)-2-aza-8,8-diethyl-8-germaspiro-4,5-decanedihydrochloride; gallium salts, and budotitane (Fig. 24).

Gallium salts
e.g. $Ga(NO_3)_3$
Galliumnitrate

$[(GeCH_2CH_2COOH)_2O_3]n$
Germanium-132
Carboxyethylgermaniumsesquioxide

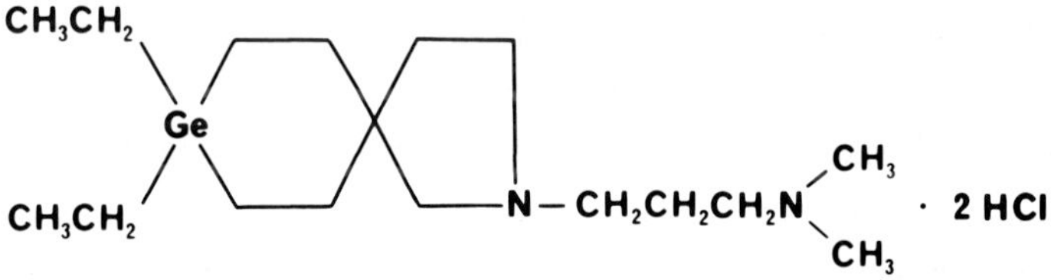

Spirogermanium
N-(3-Dimethylaminopropyl)-2-aza-8,8-diethyl-8-germaspiro-4,5-decanedihydrochloride

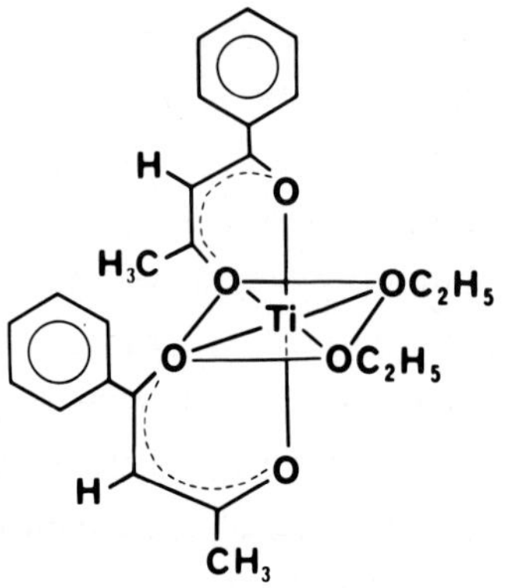

Budotitane (INN)
Diethoxybis(1-phenylbutane-1,3-dionato)titanium(IV)

Fig. 24. Four non-platinum complexes currently under clinical trials.

GALLIUM AND GERMANIUM COMPOUNDS IN CLINICAL STUDIES

Gallium salts were examined in preclinical studies along with analogous aluminum, indium, and thallium compounds from the group IIIa of the periodic table of elements. All compounds showed in vivo activity in the intraperitoneally transplanted Walker 256 carcinosarcoma when treatment was carried out intraperitoneally. All of them showed no or only little activity in the leukemias L 1210 and P 388 and in the Ehrlich ascitic carcinoma. In systemic therapy, i.e., when the substance is not injected directly into the tumor, only gallium nitrate and indium nitrate showed some activity, particularly in the intraperitoneal therapy of the subcutaneously transplanted Walker 256 carcinosarcoma. Indium nitrate, however, is much more toxic than the corresponding gallium compound. Antitumor activity of gallium salts is quite independent of the anion used. Gallium citrates are also frequently used.

Another reason why gallium salts have been studied is that radionucleides of gallium have long since been used in the clinic for szintigraphic purposes. They accumulate in particular tumors. ^{67}Gallium is used here in particular because of its accumulation in lymphomas, bone tumors, and bone metastases.

In clinical phase I studies nephrotoxicity was a dose-limiting side-effect. In addition, there were gastrointestinal side-effects and temporary hypercalcaemia. Phase II studies confirmed activity of gallium nitrate in Hodgkin and non-Hodgkin lymphomas. This could be expected from what had been found in the phase I study. Gallium salts also have a positive effect on hypercalcaemias caused by tumors. These, however, can also be treated successfully with bisphosphonates. Since many efficient drugs are available for the treatment of lymphomas today, the clinical value of gallium salts for tumor therapy seems rather limited (Hart and Adamson, 1971; Hart et al. 1971; Adamson et al. 1975; Anghileri, 1975, 1979; Anghileri et al. 1987; Schwartz and Yagoda, 1984; Collery et al. 1984; Scher et al. 1987; Ward and Taylor, 1988).

Recently an interesting effect has been found with inoperable non-small cell lung carcinomas, where gallium chloride was obviously capable of reinforcing the effect of a combination therapy with cisplatin and VP16 (Collery, 1989).

Spirogermanium, apart from showing cytotoxic activity in numerous cell cultures, has a minor effect on the leukemias L 1210 and P 388, the melanoma B 16, and the Walker 256 carcinosarcoma. The clinical phase I studies confirmed neurotoxicity as the dominant side-effect, which could be expected from preclinical experiments. There was no myelosuppression and no nephrotoxicity. Meanwhile numerous phase II studies have been carried out with different tumor types such as ovarian carcinomas, kidney tumors, glioblastomas, colon tumors, carcinomas of the prostate, malignant lymphomas, non-small cell lung carcinomas, carcinomas of the cervix, mammary carcinomas, and melanomas, but no sufficient tumor-inhibiting activity could be found which would make further clinical use recommendable (USAN, 1980; Hill et al. 1982; Slavik et al. 1982, 1983; Goodwin et al. 1987; Ward and Taylor, 1988).

A second germanium compound at present in clinical trials is carboxyethylgermanium-sesquioxide, germanium-132, which showed certain preclinical activity in ascitic hepatomas, experimental bladder tumors, and in the Lewis lung carcinoma. On account of animal experiments it is assumed that germanium-132 has a stimulating effect on the immune system and interferon-inducing properties. Germanium-132 is under clinical trials at present, especially in Japan, but so far no really promising activity has been found (Hopkins, 1980; Miyao et al. 1980; Kumano et al. 1980, 1985; Suzuki et al. 1985; Sugiya et al. 1986; Tsutsui et al. 1976; Ward and Taylor, 1988).

It is striking that there has been relatively little promising data on several transplantable tumors and none on autochthonous tumors or on a representative panel of human tumor xenografts for these three non-platinum complexes at present in clinical studies. This may be the reason why these three complexes have so far shown little promising clinical activity. The situation is quite different with the preclinical development of budotitane, the first transition metal complex besides the platinum compounds to have entered clinical studies. The development of this compound will be described somewhat more in detail in the following.

PRECLINICAL AND CLINICAL DEVELOPMENT OF BUDOTITANE

Budotitane belongs to the class of bis(ß-diketonato) metal complexes. Antitumor activity of some representatives of these complexes was reported on as early as 1982. Here we dealt with the cis-dihalogenobis(1-phenyl-1,3-butanedionato)titanium(IV) complexes with fluoride, chloride, and bromide as further ligands. These complexes effected a doubling and even a trebling of survival time of treated animals in the Walker 256 carcinosarcoma and in a transplantable murine leukemia (Keller et al. 1982, 1983).

In later studies budotitane, diethoxybis(1-phenylbutane-1,3-dionato)titanium(IV), $Ti(bzac)_2(OEt)_2$, was selected from this class of compounds for further development. In the following we will describe the most important stages in the development of this drug (Keppler et al. 1985, 1988, 1989; Keppler and Heim, 1988; Keppler and Michels, 1985; Keppler and Schmähl, 1986; Bischoff et al. 1987; Garzon et al. 1987; Heim and Keppler,

1989; Mattern et al. 1984).

The bis(ß-diketonato) metal complexes can be synthesized from the corresponding metal tetrahalogenides or tetraalkoxides and the diketonates in an anhydrous organic solvent (Fig. 25). An exception must be made for the corresponding molybdenum compounds, where the basis is molybdenumpentachloride. The compounds synthesized this way are six-coordinated, quasi-octaedrically configurated compounds, which may come either in the cis- or the trans- form (Fig. 26).

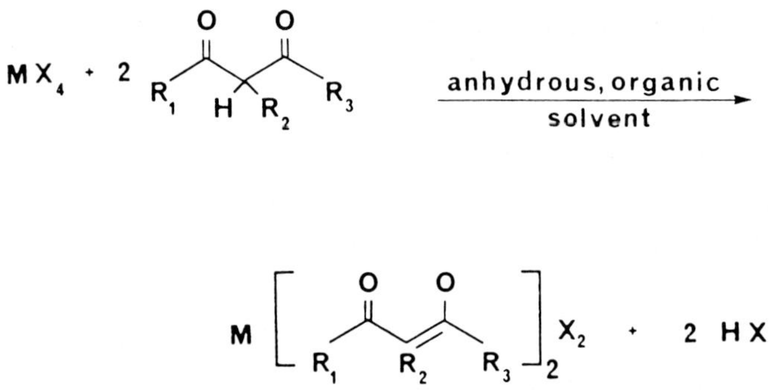

Fig. 25. General synthesis of M(ß-diketonato)$_2$X$_2$ complexes. X = Hal. or OR; M = Ti, Zr, Hf, Ge, Sn; R = organic group.

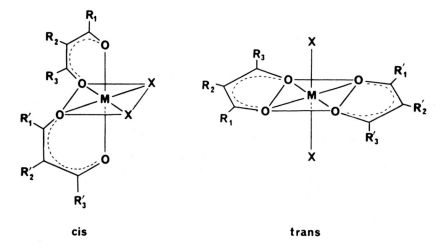

Fig. 26. Structures of the Bis(ß-diketonato) metal complexes M(ß-diketonato)$_2$X$_2$, cis- and trans- configuration.

It is surprising that the cis- configuration is usually favored, although the trans- isomer should actually be preferred for steric reasons. In the case of the benzoylacetonato complexes, which have been the centre of our investigations and which produce a maximum of antitumor activity, trans- isomers can be obtained only with extremely bulky substituents (X) such as iodide or p-dimethylaminophenoxy, as proved by means of NMR spectroscopy.

Within the cis- and the trans- form different isomers are possible. Their number depends on whether the bound diketone carries the same substituents in 1- and 3- position or different ones. At room temperature, however, these isomers of the cis- and the trans- form can convert into one another in solution, with the result that it cannot be determined which isomer is responsible for biological activity (Keppler and Heim, 1988; Keppler et al. 1988).

The $M(\beta\text{-diketonato})_2X_2$ complexes are relatively difficult to dissolve in water, and they are very susceptible to hydrolysis. Thus a specific galenic formulation had to be found for these compounds. We developed what is known as a coprecipitate, consisting of cremophorEL, propylenglycole, and the drug in the ratio of 9/1/1. It is easy to produce a micellar solution of the drug in water from this coprecipitate, which will then remain undecomposed over several hours and which is suitable for infusion therapy. This galenic formulation was used for most of the preclinical experiments and the clinical studies carried out so far (Keppler and Schmähl, 1986).

STRUCTURE-ACTIVITY RELATION OF TUMOR-INHIBITING BIS(ß-DIKETONATO) METAL COMPLEXES

The relation between chemical structure and antitumor activity of bis(ß-diketonato) metal complexes will be described in the following on the basis of some characteristic representatives chosen from about 200 complexes synthesized and examined. These representatives are presented in Tables in the order of increasing activity in the sarcoma 180 ascitic tumor model.

VARIATION OF THE (ß-DIKETONATO) LIGAND

The data in Table 4 demonstrates that the acetylacetonate ligand in the molecule of $Ti(acac)_2(OEt)_2$ is not responsible for antitumor activity (T/C = 90 - 100 %). The situation changes a little when a methyl group is replaced by a tertiary butyl group. Activity increases to T/C values of 130 - 170 %. Including other bulky substituents - in Table 4 there is a cyclohexane-substituted derivative as an example - leads to a further increase in activity up to maximum T/C values of 200 %. Sole substitution of the hydrolizable group - chloride instead of ethoxide as the last example in Table 4 - does not affect antitumor activity. It could also be confirmed in the case of other diketonates that antitumor activity is largely independent of the leaving group X. The overall trend is that substitution of the acetylacetonato ligand for lipophile, bulky alkyl or cycloalkyl groups increases efficacy of the metal complex (Keppler and Heim, 1988; Keppler and Vongerichten, 1989).

ß-diketonate	X	T/C (%)*
acetylacetonate	OEt	90 - 100
t-butyl substituted	OEt	130 - 170
t-butyl, cyclohexylmethyl substituted	OEt	150 - 200
t-butyl, cyclohexylmethyl substituted	Cl	150 - 200

Table 4. Antitumor activity of titanium complexes of the type $Ti(dik)_2X_2$ in the sarcoma 180 ascitic system at a single dose of 0.2 mmol/kg of metal complex given 24 hours after tumor transplantation. As diketonate ligand we used acetylacetone and compounds derived from it with space-filling aliphatic substituents. T/C (%) = (median survival time of treated animals vs. median survival time of control animals) x 100.

Furthermore, we examined the extent to which aromatic groups on the ligand change the efficacy of the complexes. Such data is summarized in Table 5. We compared the effect of phenyl substituents on antitumor activity of the compounds with that of Ti(acac)$_2$(OEt)$_2$, which reaches T/C values between 90 and 100 %. When a hydrogen atom of a methyl group is replaced by a phenyl group, antitumor activity will increase at about the same level as when introducing a bulky alkyl group (see Table 4). If, however, the phenyl group stands in direct conjugation to the metal enolate ring, a surprisingly high increase in antitumor activity results. For example, the compound with a phenyl group in 2- position shows T/C values between 200 and 250 %, and the derivative with a phenyl group in 1- position even reaches T/C values > 300 %.

One of the two highly active compounds is budotitane (no. 4 in the Table), which is in clinical trials today. The phenyl group of the benzoylacetonato ligand hence changes the overall structure of the metal complex in such a way that an inactive complex turns into a highly active substance.

Similarly striking effects can be obtained with the chloride derivative (no. 5 in Table 5). This confirms that the hydrolizable group has little influence on antitumor activity. However, the ethoxy group makes for much better stability as to hydrolysis. Thus the corresponding diethoxy complex is to be preferred to the chloride derivative in practice, as far as galenic formulation is concerned.

β-diketonate	X	T/C (%)*
(acetylacetonate)	OEt	90 - 100
(benzyl-acetylacetonate)	OEt	130 - 170
(2-phenyl-acetylacetonate)	OEt	200 - 250
(1-phenyl/benzoylacetonate)	OEt	> 300
(1-phenyl/benzoylacetonate)	Cl	> 300

Table 5. Antitumor activity of Ti(dik)$_2$X$_2$ complexes in the sarcoma 180 ascitic tumor at a dose of 0.2 mmol/kg given 24 hours after tumor transplantation. As diketonate ligand we used acetylacetone and corresponding phenyl-substituted derivatives. *T/C = (median survival time of treated animals vs. median survival time of control animals) x 100.

We might suppose now that systematic variations at the phenyl ring of budotitane would optimize its activity even more. We screened the influence of substituents with negative and positive inductive and mesomer effects on antitumor activity. Table 6 shows some characteristic derivatives with their T/C values. Obviously the introduction of methyl groups at the phenyl ring does not alter antitumor activity, whereas methoxy, chlorine, and nitro groups reduce antitumor activity. Thus no increase in antitumor activity can be achieved this way.

All data suggests that <u>unsubstituted</u> aromatic ring systems in the periphery of the molecule have definitely positive effects on the antitumor activity of such metal complexes, while substitutions at this site may be considered "harmful" (Keppler and Heim, 1988; Keppler and Vongerichten, 1989).

ß-diketonate	T/C (%)*
(phenyl)	> 300
(dimethylphenyl)	> 300
(3,4-dimethoxyphenyl)	150-200
(4-chlorophenyl)	150-200
(4-nitrophenyl)	100-120

Table 6. Antitumor activity of Ti(dik)$_2$Cl$_2$ complexes in the sarcoma 180 ascitic tumor at a dose of 0.2 mol/kg given 24 hours after intraperitoneal tumor transplantation. As diketonate ligand we used benzoylacetone and phenyl-substituted derivatives. T/C = (median survival time of treated animals vs. median survival time of control animals) x 100.

VARIATION OF THE CENTRAL METAL

Another important parameter in optimizing these drugs is the central metal of the complexes. The activity of the titanium and the zirconium derivatives with the benzoylacetonato ligand that produces the highest level of antitumor activity is still relatively similar, but it will then decrease markedly in the order Hf > Mo > Sn > Ge (Table 7).

M	T/C (%)*
Ti	> 300
Zr	> 300
Hf	200 - 250
Mo	150 - 200
Ge	100 - 110
Sn	120 - 150

Table 7: Antitumor activity of $M(bzac)_2Cl_2$ complexes with different central metals in the sarcoma 180 ascitic tumor at a dose of 0.2 mmol/kg, applied 24 hours after intraperitoneal tumor transplantation. *T/C = (median survival time of treated animals vs. median survival time of control animals) x 100.

While the germanium compound is virtually inactive even at other dosages, the tin compound shows an increase in activity at lower doses. T/C values around 200 % are reached with 0.1 mmol/kg. However, therapeutic range is considerably lower than in the case of the corresponding titanium compound. The molybdenum complex, which still shows some activity in the sarcoma 180, has a rather stimulating effect on tumor growth in the acetoxymethylmethylnitrosamine-induced colorectal tumors of the rat. This is quite in contrast to the excellent effect of the corresponding titanium compound on these tumor models (Garzon et al., 1987).

If the homologues zirconium and hafnium are used instead of titanium, it is also possible to synthesize hepta-coordinated compounds of the type $Zr(bzac)_3Cl$ and $Hf(bzac)_3Cl$. These are also active, even if at considerably higher doses than the corresponding octaedric coordinated complexes. It remains to be investigated whether these complexes act in the organism directly as seven-coordinated molecules or only after they have been converted into the corresponding octaedric configurated, six-coordinated complexes (Keppler and Michels, 1985; Keppler and Heim, 1988).

VARIATION OF THE GROUP X

The nature of the leaving group X does not seem to contribute much to the antitumor activity of the substance class. Ethoxide, chloride, bromide, and fluoride with benzoylacetone as ligand show excellent tumor-inhibiting activity. The same derivatives do not show any activity when they have acetylacetone as diketonate ligand (see Table 8). The four derivatives of the type $Ti(acac)_2X_2$ with X = F, Cl, Br, or OEt are inactive with T/C values between 90 and 100 %. In contrast to this, the four $Ti(bzac)_2X_2$ derivatives reach T/C values up to 300 %. They are thus highly active, irrespective of the hydrolizable group X. As mentioned before, however, galenic behaviour is considerably influenced by this leaving group, because stability in water clearly increases in the order iodine < bromine < chlorine < fluorine < OR. Thus budotitane, rather than other, analogous compounds, was chosen for further development. The iodine compound is too unstable to be considered for further development, and the bromine and fluorine compounds, apart from galenic disadvantages, have disadvantages over ethoxide as to the way in which the hydrolizable group is physiologically tolerated (Keppler and Heim, 1988).

ß-diketonate	X	T/C (%)*
acetylacetonate (CH₃COCH=C(OH)CH₃)	F	90 - 100
"	Cl	90 - 100
"	Br	90 - 100
"	OEt	90 - 100
benzoylacetonate (PhCOCH=C(OH)CH₃)	F	> 300
"	Cl	> 300
"	Br	> 300
"	OEt	> 300

Table 8. Antitumor activity of Ti(dik)$_2$X$_2$ complexes with different "leaving" groups (X) in the sarcoma 180 ascitic tumor at a dose of 0.2 mmol/kg, applied 24 hours after intraperitoneal tumor transplantation; *T/C = (median survival time of treated animals vs. median survival time of control animals) x 100.

ANTITUMOR ACTIVITY IN OTHER TRANSPLANTABLE TUMOR MODELS

Antitumor activity in other transplantable tumor models is summarized in Table 9. T/C values of > 300 % were observed in the Stockholm ascitic tumor, in the Ehrlich ascitic tumor, and in the MAC 15A colon tumor, which is a transplantable colon adenocarcinoma. T/C values of 200 % were reached in the Walker 256 carcinosarcoma. The subcutaneously transplanted sarcoma 180 can be cured with intravenous budotitane therapy. Tumor weight of the intramuscularly growing sarcoma 180 is reduced to 30 %, compared to the control experiment (100 %), given intravenous therapy.

It is interesting that this substance would not have been taken into account in a primary screening in the leukemias P 388 or L 1210, because it is only marginally active in these models (T/C = roughly 130 %). These quick-growing leukemias are certainly not the right model for finding substances that are active in slow-growing tumors such as colon tumors. However, the slow-growing tumors, of all cancers, present the biggest problem in cancer therapy today (Keppler and Schmähl, 1986; Keppler and Heim, 1988).

Tumor	Evaluation parameter	Optimum T/C-value
sarcoma 180 ascitic tumor	ST	>300
sarcoma 180 tumor, subcutaneously growing	TW	0
sarcoma 180 tumor, intramuscularly growing	TW	30
Walker 256 carcinosarcoma	ST	200
P 388 leukemia	ST	130
Stockholm ascitic tumor	ST	>300
Ehrlich Ascitic tumor	ST	>300
MAC 15A colon tumor	ST	>300

Table 9: The most important results of budotitane therapy in transplantable tumors; T/C values > 300 % mean that a high percentage of animals is cured; T/C (%) = (median tumor weight or survival time of treated animals vs. control animals) x 100; ST = median survival time; TW = median tumor weight.

THERAPY RESULTS ON AUTOCHTHONOUS, AMMN-INDUCED, COLORECTAL TUMORS WITH BUDOTITANE

The high predictivity of autochthonous tumor models for the clinical situation and, more specifically, the high predictivity of AMMN-induced colorectal tumors has already been described. Fig. 27 compares the activity of 5-fluorouracil, cisplatin, and budotitane in this model. Budotitane is markedly more active than 5-fluorouracil. It reduces tumor volume to about 20 % of the initial value. 5-Fluorouracil effects a tumor remission up to

40 % of tumor volume, whereas cisplatin, with a value of about 120 %, stimulates tumor growth a little. Stimulating effects are not infrequent with inactive compounds.

As a result of the experiments in autochthonous colon tumors, an interesting clinical indication could be observed for budotitane. This is especially important because the colon tumors are among the most frequent causes of death from cancer (Bischoff et al. 1987; Keppler and Heim, 1988).

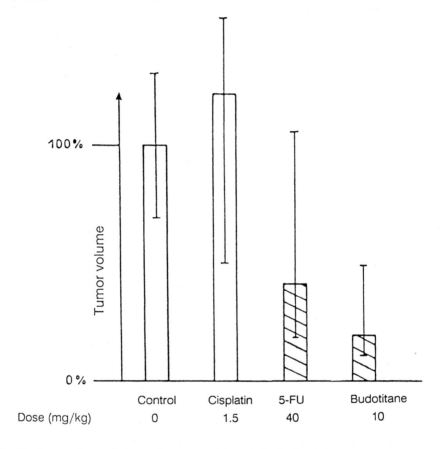

Fig. 27. Comparison of the antitumor activity of cisplatin, 5-fluorouracil, and budotitane in autochthonous, AMMN-induced colorectal tumors of the rat. The compounds listed were applied twice per week over ten weeks after tumor manifestation. The reduction of tumor volume, which is represented by the shaded columns, is statistically significant, in contrast to the control group (Kruskal Wallis Test).

TOXICITY OF BUDOTITANE

When budotitane is given to female SD rats in a single intravenous administration, the LD_{50} - that is the dose at which 50 % of the animals will die - is about 80 mg/kg. In mice the LD_{50} is about twice as high. Given i.p. application, rats and mice will tolerate somewhat more than twice the maximum tolerated i.v. dose.

In the course of these experiments we also found the major side-effects. A dose-limiting factor for budotitane is hepatotoxicity, with multiple focal necroses of the liver at about the level of the LD_{50}. There were also signs of a certain lung toxicity owing to hemorrhagic pleural effusions and hemorrhagic oedematous districts in the lung, which, however, were only found at the level of the highest lethal doses.

Chronic doses between 10 and 20 mg/kg, given twice per week over ten weeks, were tolerated without problems, with only low and reversible liver toxicity.

Laboratory parameters such as the liver enzymes of GOT and GPT, as well as LDH, were increased. In this experiment, no signs of myelosuppression could be detected in the peripheral blood.

Independent investigations in other laboratories confirmed the level of the LD_{50} to be at about 60-70 mg/kg in single intravenous administration. The other toxicological parameters were also confirmed. In addition, signs of an extramedullar blood formation were found in the liver and in the spleen. There was mild nephrotoxicity, and, in addition, alkaline phosphatase in the serum was increased.

In chronic application, up to 18 mg/kg twice a week over twelve weeks were tolerated without mortality, something which equals a total dose of 432 mg/kg, i.e., seven times the LD_{50} in single application. In this experiment we also observed a marked increase in creatinine and urea, which points to a certain degree of nephrotoxicity.

Budotitane did not cause emesis in experiments with pigeons.

Mutagenicity of budotitane was screened by means of the salmonella typhimurium mammalian microsome assay of Ames. There were no signs of a mutagenic potential.

On the whole, toxicological studies with budotitane, particularly in chronic application, show that the substance is well tolerated at levels which can be considered for therapy. Mild, reversible liver toxicity is a prominent feature. Nephrotoxicity seems to play a role only at considerably higher doses. The lack of myelosuppression is another advantage of budotitane (Keppler and Heim, 1988; Keppler et al. 1988).

BUDOTITANE CLINICAL PHASE I STUDY

In 1986 a phase I study with budotitane in cancer patients was begun. In the first part of the study only single applications were investigated, and in the second part budotitane was administered twice a week over four weeks. Seven different doses were tried out in single application, namely 1, 2, 4, 6, 9, 14, and 21 mg/kg of bodyweight.

First signs of drug toxicity appeared at 9 mg/kg, when a patient complained about an impairment of the sense of taste shortly after the infusion. This side-effect continued to be present even at higher doses, and finally a complete loss of taste was observed. This impairment, however, was entirely reversible in all cases and was present only for a few hours after budotitane infusion.

From a dose of 14 mg/kg onwards, a minor increase in liver enzymes and in lactate-dehydrogenase was observed. From a dose of 21 mg/kg onwards, a dose-limiting nephrotoxicity was found with a rise in urea and creatinine, which was in accordance with grade 2 toxicity on the basis of WHO criteria. This side-effect was also entirely reversible after some weeks, and urea and creatinine levels returned to normal. Signs of myelotoxicity could not be found at any dose.

The maximum tolerated budotitane dose thus ranges between 14 and 21 mg/kg. This equals about 30 % of the LD_{50} in animal experiments, and this is surprisingly high.

In pharmacokinetic investigations blood was taken from the patients 10 minutes, 1, 2, 8, and 24 hours as well as seven days after budotitane infusion. The titanium level in the serum and in the erythrocytes was determined by means of atom absorption spectroscopy. Given 14 mg/kg of budotitane, the highest titanium levels were found two hours after the infusion - between 2 and 5 µg/g -, and levels between 5 and 12 µg/g were found at a dose of 21 mg/kg. Titanium could be detected in the serum even after seven days and on one occasion even after four weeks. In the erythrocytes, titanium was found at a concentration between 1 and 2 µg/g.

After single application studies, side-effects of chronic budotitane application were screened. Proceeding from a maximum tolerated single dose that is slightly below 21 mg/kg, a total dose of 21 mg/kg (800 mg/m^2) was fixed for repeated applications, which was divided into 8 applications of 100 mg/m^2 twice a week over four weeks. So far three patients have been treated on the basis of this scheme without any dose-limiting toxicity occurring. It can be assumed that it is possible to considerably increase dosages, because in animal experiments, too, the maximum tolerated total dose in repeated applications was many times higher than in single applications (Keppler and Heim, 1988; Keppler et al. 1989; Heim and Keppler, 1989).

The overall picture of side-effects is one of great resemblance to that of preclinical toxicology studies. The clinical phase I study also confirmed the prediction that budotitane, unlike cisplatin, would not cause emesis. On the basis of our experiments with pigeons this was to be expected. The lack of this side-effect is of the utmost importance for the compliance of the patient. Intensive vomiting, as so often occurs with cisplatin therapy, has frequently led to patients disrupting therapy on their own account.
As had been predicted in preclinical studies, the main side-effects turned out to be liver and nephrotoxicity. Mild and reversible hepatotoxicity appears even at low doses, while nephrotoxicity cannot be observed unless the highest doses are applied. In case this side-effect should play a role in repeated applications - there has not as yet been any indication -, one might reduce it by means of a prophylactic hyperhydration, along with diuresis, comparable to the measures taken in cisplatin therapy.

A subsequent phase II study will reveal the extent to which preclinical expectations of the activity of budotitane in adenocarcinomas of the gastrointestinal tract can be confirmed.

DRUG TARGETING

As has already been described, there are three ways of proceeding when designing new anticancer agents. Apart from developing direct cisplatin derivatives and synthesizing tumor-inhibiting non-platinum complexes, there is a third method which proceeds from the assumption that linking cytotoxic structures to carrier molecules will result in a specific accumulation of antitumor agents in particular organs or tumors. This way it should be possible to achieve a rise in antitumor activity and a decline in side-effects on other organ systems. In the field of tumor-inhibiting metal complexes, real promising work has only been done where the linking of cisplatin or other, analogous, platinum structures to such carrier molecules is concerned. The synthesis of platinum compounds which are linked to hormones or to hormone-like substances in order to achieve specific affinity to hormone-receptor-positive tumors such as mammary carcinomas or tumors of the prostate is the most advanced. The compounds in Fig. 28 are two examples of platinum complexes showing affinity to hormone receptors. They were developed by Brunner, von Angerer and Schönenberger. At present they are in advanced preclinical studies. Both types of compound showed specific accumulation and effects on hormone-receptor-positive mammary tumors, both in vitro and in vivo. The relatively low solubility in water of the compounds poses a certain problem for further use (Schönenberger et al. 1984; Engel et al. 1987; Knebel and von Angerer, 1988; Knebel et al. 1989; Jennerwein et al. 1988; Karl et al. 1988; Voegeli et al. 1988).

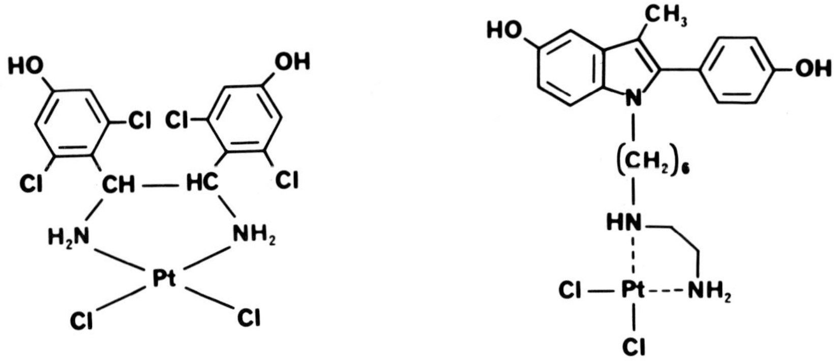

I

II

Fig. 28. Two tumor-inhibiting platinum complexes showing affinity to hormone receptors; meso-1,2-Diamino-1,2-bis(2,6-dichloro-4-hydroxyphenyl)ethanedichloro-platinum(II) (= I), 1-(2-Aminoethyl)-6,5-hydroxy-2-(4-hydroxyphenyl)-3-methylindole-1-yl-hexanedichloroplatinum(II) (= II).

Another concept of drug targeting consists in the linking of cytotoxic platinum structures to osteotropic phosphonic acids in order to achieve accumulation of platinum compounds in bone tumors and bone metastases.
With an incidence of less than 1 %, primary bone tumors are among the relatively rare neoplasms. Bone metastases are much more frequent. These can be caused by roughly half of all tumors and often present great therapeutic problems. In order to achieve a sufficiently high accumulation in the bone, cisplatin must be linked to geminal and vicinal diphosphonates in particular. These have high affinity to the hydroxylapatite, $Ca_5(OH)(PO_4)_3$, which is the inorganic component of the bone substance of

mammalians. Through chemisorption the phosphonic acids bind very strongly to the surface of the hydroxylapatite, as can be gathered from the half-life of 1-hydroxy-3-aminopropane-1,1-diphosphonic acid (APD), which is 300 days in rats (Francis and Martodam, 1983; Wingen and Schmähl, 1985).

This osteotropic property of phosphonic acids is already being used for numerous medical purposes such as technetium szintigraphy, the use of phosphonic acids in metabolic disorders of the bone, e.g., Paget's disease, and the therapy of tumor-induced hypercalcaemias, among others. The major representatives of the platinum complexes synthesized by us in this connection include the compounds AMDP and DBP (Fig. 29).

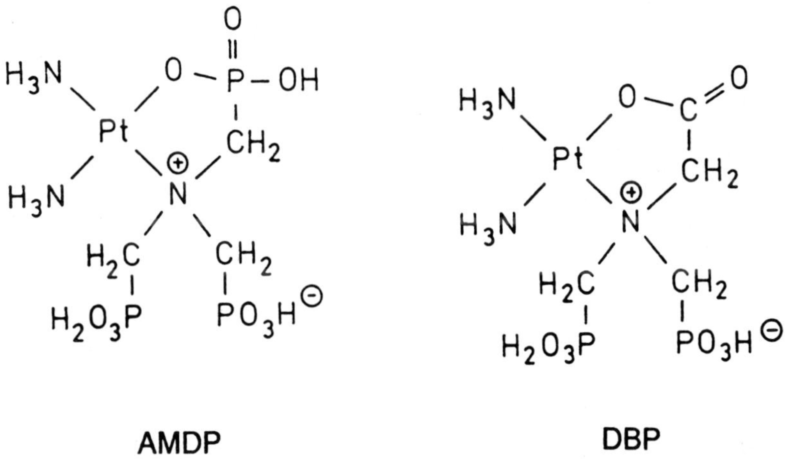

Fig. 29. cis-[Aminotris(methylenephosphonato)diamminoplatinum(II)], AMDP, and cis-[Diammino(bis(methylenephosphonato)aminoacetato)platinum(II)], DBP.

These compounds are considerably less toxic than cisplatin, as the LD_{50} in the female NMRI mouse shows. It is 310 mg/kg (0.55 mmol/kg) for AMDP, 650 mg/kg (1.25 mmol/kg) for DBP, and 12 mg/kg (0.04 mmol/kg) for cisplatin. These compounds also show marked activity in a transplantable osteosarcoma, which was initially induced by the periostal administration of ^{144}ceriumchloride to Sprague Dawley rats and later transplanted intratibially (Fig. 30). This model is relatively resistant to other chemotherapeutic agents. The compounds of AMDP and DBP effected a standstill of tumor growth during the period of therapy and also a marked prolongation of survival time.

Further experiments will have to show whether these new compounds are suited for therapy of bone tumors and bone metastases in the clinic (Klenner et al. 1988, 1989).

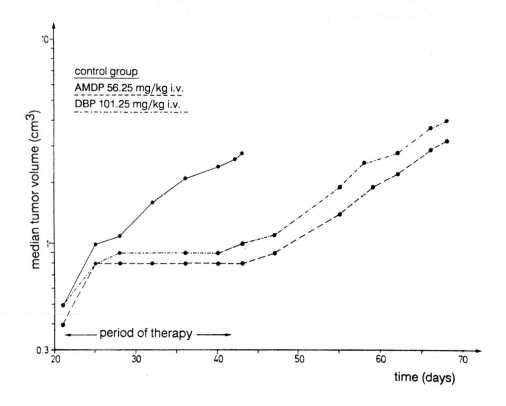

Fig. 30. Change in tumor volume during therapy of a transplantable osteosarcoma with AMDP and DBP. A standstill of tumor growth could be achieved during therapy. The untreated animals all died before day 45 after tumor transplantation, whereas some treated animals were still alive on day 65.

PERSPECTIVES

Since the turn of the century cancer has progressed from the seventh position to the second in the list of causes of death in the western world. Higher numbers are only seen with disorders of the circulatory system and the heart. Despite enormous financial and personnel efforts about 160.000 men and women die from cancer in the Federal

Republic of Germany every year. Still, we must not be too pessimistic in view of this trend, because cancer therapy makes continual progress, even if in little steps, and because with more and more tumors marked prolongations of survival time and even cures are meanwhile possible. Cancer chemotherapy, when compared to surgery and radiation, is relatively young and not as developed and optimized as the other two methods. As a consequence, we may expect the highest number of future successes of cancer chemotherapy as the youngest of the three pillars of cancer treatment. Certain childhood leukemias, Hodgkin's disease, testicular carcinomas, and some other, rather rare, tumors can be cured by chemotherapy to a high percentage today. It was precisely the possibility of being able to cure testicular carcinomas in almost any case with the help of cisplatin in combination with other measures that caused a pronounced impetus in the field of inorganic chemistry to develop more new substances, which are the subject of this article. Barnett Rosenberg, who discovered the antitumor activity of cisplatin, was an exemplary interdisciplinary research scientist. Originally a physicist, he did not hesitate to penetrate into the fields and methods of chemistry and medicine when it seemed sensible to him in connection with the development of this drug. Scientists working in this field should aim at pursueing this consistent interdisciplinary way to find new promising inorganic antitumor agents. Such promising trends can be recognized in the field of tumor-inhibiting titanium compounds, the development of budotitane, in the field of tumor-inhibiting ruthenium, gold, and tin compounds, and in drug targeting with platinum complexes showing affinity to hormone receptors and osteotropic properties.

Now as before, the development of new tumor-inhibiting compounds is associated with much preparative work. When developing new substances we are still largely dependent on empirical investigations and the setting up of empirical structure-activity relations. The extensive animal experiments, which are always carried out parallel to the syntheses, can as yet not be replaced by cell cultures or other in vitro studies, because the predictivity of a future clinical activity is only given in these models with great restrictions. In vitro studies are of importance almost only in the development of structures that are analogous to existing, well-known cytotoxic principles of action, because cytotoxic action of a new derivative as against the parent compound can usually be assessed in cell cultures.

The future will show which status the new substances from the field of inorganic chemistry will gain in cancer therapy, and it is to be hoped that this concept will make a contribution which will render possible an improvement in the therapy of cancer.

REFERENCES

Achterrath, W., Raettig, R., Franks, C.R., and Seeber, S. (1984). Aktuelle Cisplatin-Derivate. *In:* S. Seeber et al. (eds.), *Beiträge zur Onkologie,* Bd. 18, Das Resistenzproblem bei der Chemo- und Radiotherapie maligner Tumoren, 58-82, S. Karger Verlag Basel.

Adamson, R.H., Canellos, G.P., and Sieber, S.M. (1975). Studies on the Antitumor Activity of Gallium Nitrate (NSC 15200) and Other Group IIIa Metal Salts. *Cancer Chemotherapy Reports* (Part 1), 59, 3, 599-610.

Alessio, E., Attia, W., Calligaris, M., Cauci, S., Dolzani, L., Mestroni, G., Monti-Bragadin, C., Nardin, G., Quadrifoglio, F., Sava, G., Tamaro, M., and Zorzet, S. (1988). Metal Complexes of Platinum Group: The Promising Antitumor Features of cis-Dichlorotetrakis(dimethylsulfoxide)ruthenium(II) [cis-$RuCl_2(MeSO)_4$] and Related Complexes. *In:* Nicolini, M. (ed.), *Proc. of the 5th Int. Symp. on Platinum and other Metal Coordination Compounds in Cancer Chemotherapy,* 617-633, Martinus Nijhoff Publishing, Boston.

Alessio, E., Mestroni, G., Nardin, G., Attia, W.M., Calligaris, M., Sava, G., and Zorzet, S. (1988). Cis- and trans-Dihalotetrakis(dimethylsulfoxide)ruthenium(II) Complexes ($RuX_2(DMSO)_4$; X = Cl, Br): Synthesis, Structure, and Antitumor Activity. *Inorganic Chemistry,* 27, 23, 4099-4106.

Anghileri, L.J. (1975). On the Antitumor Activity of Gallium and Lanthanides. *Arzneim.-Forsch./Drug Res.* 25, 5, 793-795.

Anghileri, L.J. (1979). Effects of Gallium and Lanthanum on Experimental Tumor Growth. *Europ. J. Cancer,* 15, 1459-1462.

Anghileri, L.J., Crone-Escanye, M.-Chr., and Robert, J. (1987). Antitumor Activity of Gallium and Lanthanum: Role of Cation-Cell Membrane Interaction. *Anticancer Res.,* 7, 1205-1208.

Berger, M.R., Bischoff, H., Garzon, F.T., and Schmähl, D. (1986). Autochthonous, Acetoxymethylmethylnitrosamine-induced Colorectal Cancer in Rats: A Useful Tool in Selecting New Active Antineoplastic Agents? *Hepatogastroenterol.* 33, 227-234.

Berger, M.R., Garzon, F.T., Keppler, B.K., and Schmähl, D. (1989). Efficacy of New Ruthenium Complexes against Chemically Induced Autochthonous Colorectal Carcinoma in Rats. *Anticancer Res.,* -in press-.

Berners-Price, S.J., and Sadler, P.J. (1985 [Publ. 1986]). Gold Drugs. *Front. Bioinorg. Chem.,* 376-88.

Berners-Price, S.J., Mirabelli, Ch.K., Johnson, R.K., Mattern, M.R., McCabe, F.L., Faucette, L.F., Chiu-Mei Sung, Shau-Ming Mong, Sadler, P.J., and Crooke, St.T. (1986). In Vivo Antitumor Activity and in Vitro Cytotoxic Properties of Bis[1,2-bis-(diphenylphosphino)ethane]gold(I)chloride. *Cancer Research,* 46, 5486-5493.

Berners-Price, S.J., and Sadler, P.J. (1988). Phosphines and Metal Phosphine Complexes: Relationship of Chemistry to Anticancer and Other Biological Activity. *Structure and Bonding,* 70, 28-97.

Bischoff, H., Berger, M.R., Keppler, B.K., and Schmähl, D. (1987). Efficacy of β-Diketonato Complexes of Titanium, Zirconium, and Hafnium against Autochthonous Colonic Tumors in Rats. *J. Cancer Res. Clin. Oncol.* 113, 446-450.

Clarke, M.J. (1980). Oncological Implications of the Chemistry of Ruthenium. *In:* H. Sigel (ed.), *Metal Ions in Biological Systems,* Vol. 11: Metal Complexes as Anticancer Agents, 231-276, Marcel Dekker, New York.

Clarke, M.J. (1980). The Potential of Ruthenium in Anticancer Pharmaceuticals. *Acs. Symp. Ser.*(Am. Chem. Soc.) 140, 157-180.

Clarke, M.J., Galang, R.D., Rodriguez, V.M., Kumar, R., Pell, S., Bryan, D.M. (1988). Chemical Considerations in the Design of Ruthenium Anticancer Agents. *In:* Nicolini, M. (ed.), *Proc. of the 5th Int. Symp. on Platinum and other Metal Coordination Compounds in Cancer Chemotherapy,* 582-600, Martinus Nijhoff Publishing, Boston.

Collery, P., Millart, H., Simoneau, J.P., Pluot, M., Halpern, S., Pechery, C., Choisy, H., and Etienne, J.C. (1984). Experimental Treatment of Mammary Carcinomas by Gallium Chloride after Oral Administration: Intratumor dosages of gallium, anatomopathologic study and intracellular microanalysis. *Trace Elements in Medicine,* 1, 4, 159-161.

Collery, P. (1989). *Personal Communication.*

Curt, G.A., Allegra, C.J., Fine, R.L., Mujagic, H., Chao Yeh, G., and Chabner B.A. (1986). Cancer Chemotherapy. *In:* Ullmann's Encyclopedia of Industrial Chemistry, Vol. A5, 1-28, VCH Verlag Weinheim, FRG.

Dimitrov, N.V., and Eastland, G.W. (1978). Antitumor Effect of Rhenium Carboxylates in Tumor-Bearing Mice. *Int. Congr. Chemother., Proc. of the 10th, Current Chemother. 1977,* 1319-1321.

Eastland, G.W., Yang, G., and Thompson, T. (1983). Studies of Rhenium Carboxylates as Antitumor Agents. Part II. Antitumor Studies of Bis(μ-Propionato)Diaquotetrabromodirhenium(III) in Tumor-Bearing Mice. *Meth. and Find. Exptl. Clin. Pharmacol.,* 5 (7), 435-438.

Ehninger, G., Haag, C., and Wilms, K. (1984). Die Pharmakokinetik von cis-Diaminodichloroplatin. *TumorDiagnostik & Therapie,* 5, 147-151.

Elo, H.O., and Lumme, P.O. (1985). Antitumor Activity of trans-Bis(salicylaldoximato)-copper(II): A Novel Antiproliferative Metal Complex. *Cancer Treatment Rep.,* 69, 9, 1021-1022.

Engel, J., Schönenberger, H., Lux, F., and Hilgard, P. (1987). Estrophilic Cisplatin Derivatives. *Cancer Treatment Reviews,* 14, 275-283.

Erck, A., Rainen, L., Whileyman, J., Chang, J.M., Kimball, A.P., Bear, J. (1974). Studies of Rhodium(II) Carboxylates as Potential Antitumor Agents. *Proc. Soc. Exp. Biol. and Med.,* 145, 1278-1283.

Francis, M.D., and Martodam, R.R. (1983). Chemical, Biochemical, and Medicinal Properties of the Diphosphonates. *In:* Hilderbrand, R.L. (ed.), *The Role of Phosphonates in Living Systems.* CRC Press, 55-96.

Garzon, F.T., Berger, M.R., Keppler, B.K., and Schmähl, D. (1987). Paradoxical Effect of Dichlorobis(1-phenylbutane-1,3-dionato)molybdenum(IV), Mo(bzac)$_2$Cl$_2$, on the Growth of Autochthonous Chemically Induced Colorectal Tumors in SD Rats. *Cancer Letters,* 34, 325-330.

Garzon, F.T., Berger, M.R., Keppler, B.K., and Schmähl, D. (1987). Comparative Antitumor Activity of Ruthenium Derivatives with 5'-Deoxy-5-fluorouridine in Chemically Induced Colorectal Tumors in Sd Rats. *Cancer Chemotherapy and Pharmacology,* 19, 347-349.

Garzon, F.T., Berger, M.R., Keppler, B.K., and Schmähl, D. (1987). Activity of Heterocyclic Coordinated Ruthenium Derivatives on Experimental Acetoxymethylmethylnitrosamine-induced Colorectal Tumors in SD Rats. 5th NCI-EORTC Symposium on New Drugs in Cancer Therapy, Amsterdam, 22.-24.10.1986, *Invest. New Drugs,* 5, 1, 84.

Gill, D.S. (1984). Structure Activity Relationship of Antitumor Palladium Complexes. *Dev. Oncol.* 17, 267-278.

Giraldi, T., Zassinovich, G., and Mestroni, G. (1974). Antitumor Action of Planar, Organometallic Rhodium(I) Complexes. *Chem.-Biol. Interactions* 9, 389-394.

Giraldi, T., Sava, G., Bertoli, G., Mestroni, G., and Zassinovich, G. (1977). Antitumor Action of Two Rhodium and Ruthenium Complexes in Comparison with cis-Diamminedichloroplatinum(II). *Cancer Res.* 37, 2662-2666.

Goodwin, J.W., Kopecky, K., Slavik, M., Tranum, B.L., Balcerzak, St.P., Fletcher, W.S., and Costanzi, J.J. (1987). Phase II Evaluation of Spirogermanium in Malignant Melanoma: A Southwest Oncology Group Study. *Cancer Treatment Rep.,* 71, 10, 985-986.

Harrap, K.R. (1985). Preclinical Studies Identifying Carboplatin as a Viable Cisplatin Alternative. *Cancer Treatment Rev.,* 12 (Suppl. A), 21-33.

Hart, M.M., and Adamson, R.H. (1971). Antitumor Activity and Toxicity of Salts of Inorganic Group IIIa Metals: Aluminum, Gallium, Indium, and Thallium. *Proc. Nat. Acad. Sci. USA,* 68, 7, 1623-1626.

Hart, M.M., Smith, C.F., Yancey, S.T., and Adamson, R.H. (1971). Toxicity and Antitumor Activity of Gallium Nitrate and Periodically Related Metal Salts. *Journal of the National Cancer Institute,* 47, 5, 1121-1127.

Heim, M.E., and Keppler, B.K. (1989). Clinical Studies with Budotitane - A New non-Platinum Metal Complex for Cancer Therapy. *Progress in Clin. Biochemistry and Medicine,* 10, 217-223.

Hill, B.T., Whatley, S.A., Bellamy, A.S., Jenkins, L.Y., and Whelan, R.D.H. (1982). Cytotoxic Effects and Biological Activity of 2-Aza-8-germanspiro[4,5]decane-2-propanamine-8,8-diethyl-N,N-dimethyl Dichloride (NSC 192965; Spirogermanium) in Vitro. *Cancer Res.,* 42, 2852-2856.

Hodnett, E.M., Moore, Ch.H., and French, F.A. (1971). Cobalt Chelates of Schiff Bases of Aromatic Amines as Antitumor Agents. *J. Medicinal Chem.,* 14, 11, 1121-1123.

Hopkins, S.J. (1980). Ge-132. *Drugs of the Future,* V, 11, 545-546.

Howard, R.A., Sherwood, E., Erck, A., Kimball, A.P., Bear, J.L. (1977). Hydrophobicity of Several Rhodium(II) Carboxylates Correlated with Their Biologic Activity. *J. Medicinal Chem.,* 20, 7, 943-946.

Jennerwein, M., Wappes, B., Gust, R., Schönenberger, H., Engel, J., Seeber, S., and Osieka, R. (1988). Influence of Ring Substituents on the Antitumor Effect of Dichloro(1,2-diphenylethylenediamine)platinum(II) Complexes. *J. Cancer Res. Clin. Oncol.,* 114, 347-358.

Karl, J., Gust, R., and Spruss, Th.(1988). Ring-Substituted [1,2-Bis(4-hydroxyphenyl)-ethylenediamine]dichloroplatinum(II) Complexes: Comparison with a Selective Effect on the Hormone-Dependent Mammary Carcinoma. *J. Medicinal Chem.,* 31, 72-83.

Keller, H.J., Keppler, B.K., and Schmähl, D. (1982). Antitumor Activity of cis-Dihalogenobis(1-phenyl-1,3-dionato)titanium(IV) Compounds against Walker 256 Carcinosarcoma. *Arzneim.-Forsch./Drug Res.* 32 (II), 8, 806-807.

Keller, H.J., Keppler, B.K., and Schmähl, D. (1983). Antitumor Activity of cis-Dihalogenobis(1-phenyl-1,3-dionato)titanium(IV) Compounds. *J. Cancer Res. Clin. Oncol.* 105, 109-110.

Kempf, S.R., and Ivankovic, S. (1986). Carcinogenic Effect of Cisplatin (cis-Di-amminedichloroplatinum(II), CDDP) in BD IX Rats. *J. Cancer Res. Clin. Oncol.,* 111, 133-136.

Kempf, S.R., and Ivankovic, S. (1986). Chemotherapy-Induced Malignancies in Rats after Treatment with Cisplatin as a Single Agent and in Combination: Preliminary Studies. *Oncology,* 43, 187-191.

Keppler, B.K., and Michels, K. (1985). Antitumor Activity of 1,3-Diketonato Zirconium(IV) and Hafnium(IV) Complexes. *Arzneim.-Forsch./Drug Res.* 35 (II), 12, 1837-1839.

Keppler, B.K., Diez, A., and Seifried, V. (1985). Antitumor Activity of Phenyl Substituted Dihalogenobis(1-phenyl-1,3-butanedionato)titanium(IV) Complexes. *Arzneim.-Forsch./Drug Res.* 35 (II), 12, 1832-1836.

Keppler, B.K., and Rupp, W. (1986). Antitumor Activity of Imidazolium-bis(imidazole)-tetrachlororuthenate(III). *J. Cancer Res. Clin. Oncol.* 111, 166-168.

Keppler, B.K., and Schmähl, D. (1986). Preclinical Evaluation of Dichlorobis(1-phenyl-butane-1,3-dionato)titanium(IV) and Budotitane. *Arzneim.-Forsch./Drug Res.* 36 (II), 12, 1822-1828.

Keppler, B.K. (1987). Metallkomplexe in der Krebstherapie. *Nachr. Chem. Tech. Lab.,* 35, 10, 1029-1036.

Keppler, B.K., Balzer, W., and Seifried, V. (1987). Synthesis and Antitumor Activity of Triazolium-bis(triazole)tetrachlororuthenate(III) and Bistriazolium-triazolepentachlororuthenate(III). *Arzneim.-Forsch./Drug Res.* 37(II), 7, 770-771.

Keppler, B.K., Wehe, D., Endres, H., and Rupp, W. (1987). Synthesis, Antitumor Activity, and X-Ray Structure of Bis(imidazolium)imidazolepentachlororuthenate(III), $(ImH)_2(RuImCl_5)$. *Inorganic Chemistry,* 26 (6), 844-846.

Keppler, B.K., Rupp, W., Endres, H., Niebl, R., and Balzer, W. (1987). Synthesis, Molecular Structure, and Tumor-Inhibiting Properties of Imidazolium-bis(imidazole)-tetrachlororuthenate(III) and its Methyl-Substituted Derivatives. *Inorganic Chemistry*, 26, 4366-4370.

Keppler, B.K., Garzon, F.T., Rupp, W., Niebl, R., Juhl, U.M., Berger, M.R., and Schmähl, D. (1987). Preclinical Evaluation of New Tumor-Inhibiting Ruthenium Compounds. *Proc. 4th SEK-Symp.*, Heidelberg, 18.-21.3.1987, *J. Cancer Res. Clin. Oncol.*, Suppl. to Vol. 113.

Keppler, B.K., Bischoff, H., Berger, M.R., Heim, M.E., Reznik, G., and Schmähl, D. (1988). Preclinical Development and First Clinical Studies of Budotitane. ISPCC 1987, Padua; *In:* Nicolini, M. (Ed.), *Proc. 5th Int. Symp. on Platinum and other Metal Coordination Complexes in Cancer Chemotherapy*, Martinus Nijhoff Publishing, Boston, 684-694.

Keppler, B.K., and Heim, M.E. (1988). Antitumor-Active Bis-β-Diketonato Metal Complexes: Budotitane - A New Anticancer Agent. *Drugs of the Future*, 13, 5-6, 637-652.

Keppler, B.K., Henn, M., Juhl, U.M., Berger, M.R., Niebl, R.E., and Wagner, F.E. (1989). New Ruthenium Complexes for the Treatment of Cancer. *Progress in Clinical Biochemistry and Medicine*, 10, 41-70.

Keppler, B.K., Heim, M.E., Flechtner, H., Wingen, F., and Pool, B.L. (1989). Assessment of the Antitumor Activity of Budotitane in Three Different Transplantable Tumor Models, its Lack of Mutagenicity, and First Results of Clinical Phase I Studies. *Arzneim.-Forsch./Drug Res.* 39 (I), 6, 706-709.

Keppler, B.K., and Vongerichten, H. (1989). *-unpublished results-*.

Klenner, T., Keppler, B.K., Amelung, F., and Schmähl, D. (1989). Aminotris(methylenephosphonato)diaminoplatinum(II) [AMDP], a New Anticancer Agent Superior to Cisplatin (CDDP) in the Transplantable Rat Osteosarcoma. 5. SEK-Symposium, Heidelberg, 10.-12.4.1989, *Suppl. J. Cancer Res. Clin. Oncol.* 115, TH 5.

Klenner, T., Münch, H., Wingen, F., Schmähl, D., and Keppler, B.K. (1988). Efficacy of New Cisplatin-linked Bisphosphonates in Transplantable Rat Osteosarcoma. *Proc. 19th National Cancer Congress*, Frankfurt, 28.2.-5.3.1988, *J. Cancer Res. Clin. Oncol.*, Suppl. to Vol. 114.

Knebel, N., and von Angerer, E. (1988). Platinum Complexes with Binding Affinity for the Estrogen Receptor. *J. Medicinal Chem.*, 31, 1675-1679.

Knebel, N., Schiller, Cl.-D., Schneider, M.R., Schönenberger, H., and von Angerer, E. (1989). Carrier Mediated Action of Platinum Complexes on Estrogen Receptor Positive Tumors. *Eur. J. Cancer Clin. Oncol.*, 25, 2, 293-299.

Kociba, R.J., Sleight, S.D., and Rosenberg, B. (1970). Inhibition of Dunning Ascitic Leukemia and Walker 256 Carcinosarcoma with cis-Diamminedichloroplatinum (NSC 119875). *Cancer Chemotherapy Reports* (Part 1), 54, 5, 325-328.

Köpf-Maier, P., and Köpf, H. (1988). Transition and Main-Group Metal Cyclopentadienyl Complexes: Preclinical Studies on a Series of Antitumor Agents of Different Structural Type. *Structure and Bonding*, 70, 105-181.

Köpf-Maier, P., and Köpf, H. (1988). Antitumor Cyclopentadienyl Metal Complexes: Current Status and Recent Pharmacological Results. *In:* Gielen, M.F. (ed.), *Metal-Based Anti-tumour Drugs,* Freund Publishing House, London, 55-102.

Krakoff, I.H. (1988). The Development of More Effective Platinum Therapy. *In:* M. Nicolini (ed.), *Proc. of the 5th Int. Symp. on Platinum and other Metal Coordination Compounds in Cancer Chemotherapy,* Martinus Nijhoff Publishing, Boston, 351-354.

Kumano, N., Nakai, Y., Ishikawa, T., Koinumaru, S., Suzuki, S., Kikumoto, T., and Konno, K. (1980). Antitumor Effect of Organogermanium Compound (Ge-132) in Mouse Tumors. *In:* Nelson, J.D., and Grassi, C. (eds.), *Current Chemotherapy and Infectious Disease,* Proc. Int. Congr. Chemother. 11th, 1979, Am. Soc. Microbiol., Washington, 1525-1527.

Kumano, N., Ishikawa, T., Koinumaru, S., Kikumoto, T., Suzuki, S., Nakai, Y., and Konno, K. (1985). Antitumor Effect of the Organogermanium Compound Ge-132 on the Lewis Lung Carcinoma (3LL) in C57BL/6 (B6) Mice. *Tohuku J. Exp. Med.,* 146, 97-104.

Leopold, W.R., Miller, E.C., and Miller, J.A. (1979). Carcinogenicity of Antitumor cis-Platinum(II) Coordination Complexes in the Mouse and Rat. *Cancer Res.,* 39, 913-918.

Lewis, A.J., and Walz, D.T. (1982). Immunopharmacology of Gold. *In:* G.P. Ellis, G.B. West (eds.), *Progress in Medicinal Chem.,* 19, Elsevier Biomedical Press, 2-49.

Lippard, St.J. (1981). Binding of the Antitumor Drug cis-Diamminedichloroplatinum(II) to DNA and to the Nucleosome Core Particle. *In:* Ramaswamy H. Sarma (ed.), *Biomolecular Stereodynamics,* Vol. II, 165-183, Adenine Press, New York.

Litterst, Ch.L., LeRoy, A.F., Guarino, A.M. (1979). Disposition and Distribution of Platinum Following Parental Administration of cis-Dichlorodiammineplatinum(II) to Animals. *Cancer Treatment Rep.,* 63, 9-10, 1485-1492.

Lumme, P., Elo, H., and Jänne, J. (1984). Antitumor Activity and Metal Complexes of the First Transition Series. *Trans*-bis(salicylaldoximato)copper(II) and Related Copper(II) Complexes, a Novel Group of Potential Antitumor Agents. *Inorganica Chimica Acta,* 92, 241-251.

Lumme, P.O., and Elo, H.O. (1985). Antitumor Activity and Metal Complexes, a Comparison. *Inorganica Chimica Acta,* 107, L15-L16.

Mattern, J., Keppler, B.K., and Volm, M. (1984). Preclinical Evaluation of Diethoxy(1-phenyl-1,3-dionato)titanium(IV) in Human Tumor Xenografts. *Arzneim.-Forsch./Drug Res.* 34 (II), 10, 1289-1290.

Miyao, K., Onishi, T., Asai, K., Tomizawa, S., and Suzuki, F. (1980). Toxicology and Phase I Studies on a Novel Organogermanium Compound, Ge-132. *In:* Nelson, J.D., and Grassi, C. (eds.), *Current Chemotherapy and Infectious Disease,* Vol. II, 1527-1529.

Peyrone, M. (1844). Über die Einwirkung des Ammoniak auf Platinchlorür. *Annalen der Chemie und Pharmacie,* LI, 1 ff.

Pinto, A.L., and Lippard, St.J. (1985). Binding of the Antitumor Drug cis-Diamminedichloroplatinum(II) (Cisplatin) to DNA. *Biochimica et Biophysica Acta,* 780, 167-180.

Pöldinger, W. (1982). Kompendium der Psychopharmakotherapie. *Editiones "Roche"*, Basel, 126.

Prestayko, A.W. (1981). Clinical Pharmacology of Cisplatin. *Cancer and Chemotherapy*, III, 351-356.

Reedijk, J. (1987). The mechanism of action of platinum anti-tumor drugs. *Pure Appl. Chem.*, 59 (2), 181-192.

Rose, W.C., and Schurig, J.E. (1985). Preclinical Antitumor and Toxicologic Profile of Carboplatin. *Cancer Treatment Rev.*, 12 (Suppl. A), 1-19.

Rosenberg, B., and VanCamp, L. (1969). Platinum Compounds: A New Class of Potent Antitumor Agents. *Nature*, 222, 385-386.

Rosenberg, B., and VanCamp, L. (1970). The Successful Regression of Large Solid Sarcoma 180 Tumors by Platinum Compounds. *Cancer Res.*, 304, 1799-1802.

Rosenberg, B. (1975). Possible Mechanisms for the Antitumor Activity of Platinum Coordination Complexes. *Cancer Chemotherapy Rep.* (Part 1), 59, 3, 589-598.

Rosenberg, B. (1978). Platinum Complexes for the Treatment of Cancer. *Interdisciplinary Science Reviews*, 3, 2, 134-147.

Rosenberg, B. (1978). Platinum Complex - DNA Interactions and Anticancer Activity. *Biochemie*, 60, 859-867.

Sadler, P.J., Nasr, M., and Narayanan, V.L. (1984). The Design of Metal Complexes as Anticancer Agents. *Proc. of the 4th Int. Symp. on Platinum Coordination Complexes in Cancer Chemotherapy*, 290-304, Martinus Nijhoff Publishing, Boston.

Sava, G., Giraldi, T., Mestroni, G., and Zassinovich, G. (1983). Antitumor Effects of Rhodium(I), Iridium (I), and Ruthenium(II) Complexes in Comparison with cis-Dichlorodiamminoplatinum(II). *Chem.-Biol. Interactions*, 45, 1-6.

Sava, G., Zorzet, S., Giraldi, T., Mestroni, G., and Zassinovich, G. (1984). Antineoplastic Activity and Toxicity of an Organometallic Complex of Ruthenium(II) in Comparison with cis-PDD in Mice Bearing Solid Malignant Neoplasms. *Eur. J. Cancer Clin. Oncol.*, 20, 6, 841-847.

Sava, G., Zorzet, S., Mestroni, G., and Zassinovich, G. (1985). Antineoplastic Activity of Planar Rhodium(I) Complexes in Mice Bearing Lewis Lung Carcinoma and P 388 Leukemia. *Anticancer Res.* 5, 249-252.

Scher, H.J., Curley, T., Geller, N., Dershaw, D., Chan, E., Nisselbaum, J., Alcock, N., Hollander, P., and Yagoda, A. (1987). Gallium Nitrate in Prostatic Cancer: Evaluation of Antitumor Activity and Effects on Bone Turnover. *Cancer Treatment Rep.*, 71, 10, 887-893.

Schmähl, D., and Berger, M.R. (1988). Possibilities and Limitations of Antineoplastic Chemotherapy: Experimental and Clinical Aspects. *Int. J. Exp. Clin. Chemother.* 1, 1-11.

Schönenberger, H., Wappes, B., Jennerwein, M., and Berger, M. (1984). Entwicklung selektiv wirkender Platinkomplexe. *In:* S. Seeber et al. (eds.), Beiträge zur Onkologie, Bd. 18, 48-57, S. Karger Verlag Basel.

Schwartz, S., and Yagoda, A. (1984). Phase I-II Trial of Gallium Nitrate for Advanced Hypernephroma. *Anticancer Res.,* 4, 317-318.

Sherman, S.E., Gibson, D., Wang, A.H.-J., and Lippard, St.J. (1985). X-Ray Structure of the Major Adduct of Anticancer Drug Cisplatin with DNA: cis-[Pt(NH$_3$)$_2$(d(pGpG)]. *Science,* 230, 412-417.

Sherman, S.E., and Lippard, St.J. (1987). Structural Aspects of Anticancer Drug Interactions with DNA. *Chem. Rev.,* 87, 1153-1181.

Simon, T.M., Kunishima, D.H., Vibert, G.J., and Lorber, A. (1981). Screening Trial with the Coordinated Gold Compound Auranofin Using Mouse Lymphocytic Leukemia P 388. *Cancer Research,* 41, 94-97.

Slavik, M., Elias, L., Mrema, J., and Saiers, J.H. (1982). Laboratory and Clinical Studies of Spirogermanium, a Novel Heterocyclic Anticancer Drug. *Drugs Exptl. Clin. Res.* VIII (4), 379-385.

Slavik, M., Blanc, O., and Davis, J. (1983). Spirogermanium: A New Investigational Drug of Novel Structure and Lack of Bone Marrow Toxicity. *Invest. New Drugs,* 1, 225-234.

Sternberg, C., Cheng, E., and Sordillo, P. (1984). Phase II Trial of 1,2-Diamino-cyclohexane-(4-carboxyphthalato)platinum(II) (DACCP) in Colorectal Carcinoma. *Am. J. Clin. Oncol.* (CCT), 7, 503-505.

Sugiya, Y., Sugita, T., Sakamaki, S., Abo, Y., and Satoh, H. (1986). Subacute and Chronic Intraperitoneal Toxicity of Carboxyethylgermaniumsesquioxide (Ge-132) in Rats. *Oyo Yakuri,* 32 (1), 93-111.

Suzuki, F., Brutkiewicz, R.R., and Pollard, R.B. (1985). Ability of Sera from Mice Treated with Ge-132, an Organic Germanium Compound, to Inhibit Experimental Murine Ascites Tumours. *Br. J. Cancer,* 757-763.

Suzuki, F., Brutkiewicz, R.R., and Pollard, R.B. (1985). Importance of T-Cells and Macrophages in the Antitumour Activity of Carboxyethylgermanium Sesquioxide (Ge-132). *Anticancer Res.,* 5, 479-484.

Tsuruo, T., Iida, H., Tsukagoshi, S., and Sakurai, Y. (1980). Growth Inhibition of Lewis Lung Carcinoma by an Inorganic Dye, Ruthenium Red. *Gann,* 71, 151-154.

Tsutsui, M., Kakimoto, N., Axtell, D.D., Oikawa, H., and Asai, K. (1976). Crystal Structure of Carboxyethylgermanium Sesquioxide. *J. Am. Chem. Soc.,* 98, 25, 8287-8289.

USAN (1980). Spirogermanium Hydrochloride. *Drugs of the Future,* V, 3, 149-151.

Vermorken, J.B., ten Bokkel Huinink, W.W., McVie, J.G., van der Vijgh, W.J.F., and Pinedo, H.M. (1984). Clinical Experience with 1,1-Diaminomethylcyclohexane (Sulfato) Platinum(II) (TNO-6). *Dev. Oncol.,* 17, 330-343.

Vermorken, J.B., Winograd, B., van der Vijgh, W.J.F. (1985). Clinical Pharmacology of Cisplatin and Some New Platinum Analogs. *Recent Adv. Chemother.,* Proc. Int. Congr. Chemother., 14th, 96-99.

Voegeli, R., Pohl, J., Hilgard, P., Engel, J., Schumacher, W., Brunner, H., Schmidt, M., Holzinger, U., and Schönenberger, H. (1988). Synthesis and Therapeutic Effect of New cis-Platinum Complexes on Experimental Tumors. *In:* Nicolini, M. (ed.), *Proc. of the 5th Int. Symp. on Platinum and other Metal Coordination Compounds in Cancer Chemotherapy,* Martinus Nijhoff Publishing, Boston, 343-350.

von Heyden, H.W., Weinstock, N., Schaper, R., Beyer, J.-H., Nagel, G.A., and Seidel, D. (1980). Platinkinetik: Literaturübersicht und erste eigene Ergebnisse. *In:* S. Seeber et al. (eds.), *Beiträge zur Onkologie,* Band 3, S. Karger Verlag Basel.

Ward, S.G., and Taylor, R.C. (1988). Anti-Tumor Activity of the Main-Group Metallic Elements: Aluminum, Gallium, Indium, Thallium, Germanium, Lead, Antimony, and Bismuth. *In:* Gielen, M.F. (ed.), *Metal-Based Anti-Tumour Drugs,* Freund Publishing House, London, 1-54.

Wingen, F., and Schmähl, D. (1985). Distribution of 3-Amino-1-hydroxypropane-1,1-diphosphonic Acid in Rats and Effects on Rat Osteosarcoma. *Arzneim.-Forsch./Drug Res.* 35 (II), 10, 1565-1571.

Zeller, W.J., and Berger, M.R. (1984). Chemically Induced Autochthonous Tumor Models in Experimental Chemotherapy. *Behring Inst. Mitt.* 74, 201-208.

TIN COMPOUNDS AND THEIR POTENTIAL AS PHARMACEUTICAL AGENTS

Alan J. Crowe
International Tin Research Institute
Kingston Lane
Uxbridge UB8 3PJ
UK.

INTRODUCTION

Tin compounds may be divided into two main classes; i) inorganic tin salts in which tin, with a valency of +2 or +4, is chemically associated with an element other than carbon, or with an ionic radical; ii) organotin compounds, which possess one or more direct tin-carbon bond(s). Inorganic tin chemicals have been used for many centuries in coloured pigments for ceramics and paints, and in glazes. In contrast the organotins, which were discovered about 150 years ago, were only commercialised during the 1950's. Both classes of compounds are now widely employed in industry and show a remarkable diversity of applications (Blunden et al.1985; Karpel and Evans 1985).

With regard to the chemotherapeutic properties of tin compounds, only certain inorganic derivatives are commercially available for the treatment of humans. Tin(II) fluoride has, for many years, been used in toothpastes, dentifrices, topical solutions, mouth washes and occasionally as a constituent of dental cements of the zinc oxide-polyacrylic acid type (Blunden et al.1985). The main advantages of tin(II) fluoride for such applications are that it is more effective than sodium fluoride for the protection of dental enamel and for the reduction of dental mottling fluorosis (Muhler and van Huysen 1947; Howell et al 1955). Tin(II) fluoride is also highly effective in the control of root hypersensitivity, on its own, or with acidulated phosphate fluoride (APF). Formulations containing these two substances have been suggested to be suitable for use in the control of root caries and hypersensitivity (Shannan and Wightman 1970; Shannon 1971; Williams et al 1974). In addition to preventing dental caries, SnF_2 is greatly superior to all other fluorides in its inhibition of dental plaque (Konig 1959; Svatun and Attramadal 1978). The Sn^{2+} ion has been identified as the agent of greatest importance in plaque suppression and that the fluoride ion has little, if any, inhibitory effect (Scherer 1981). A major disadvantage associated with the use of tin(II) fluoride in dental formulations has been the instability of its aqueous solutions towards hydrolysis and oxidation. However, the addition of suitable stabilisers such as glycerol, sugars and gums has overcome this problem (Hefferren 1963; Lim 1970). Stable organic solutions of tin(II) salts have also been produced by the addition of suitable complexing agents (Blunden et al 1983). Thus the recent decline in the use of tin(II) fluoride in dental formulations may be reversed now that both its aqueous and organic solutions can be stabilised against hydrolysis and oxidation.

Dibutyltin dilaurate and tin(II) octoate are used as catalysts for the crosslinking of room temperature vulcanising (RTV) silicone rubbers, some of which are used as dental impression mouldings, maxillo-facial prosthesis materials, and as soft lining material for dentures (Blunden et al 1985). The use of dibutyltin dilaurate in such silicone rubbers has the additional benefit of inhibiting the inter-oral growth of <u>Candida albicans</u> which often occurs with these materials (Wright 1980).

Tin(II) salts are also used in radiopharmaceuticals, 99m-technetium (^{99m}Tc) is the radioisotope which has the optimum nuclear properties for clinical X-ray scanning and many imaging procedures based on this nuclide are currently in use (Keyes et al 1973). Typically the ^{99m}Tc radiopharmaceuticals are prepared by the reduction of the technetate(VII) ion ($^{99}TcO_4^-$) in the presence of a suitable ligand. Many tin(II) salts have been used for this reduction, but the preferred reductant is tin(II) chloride in aqueous hydrochloric acid (Deutsch et al 1976). A number of (^{99m}Tc-Sn)-labelled radiopharmaceuticals have been developed and many of these are used routinely in diagnostic medicine Table 1.

A pharmaceutical preparation, Stannoxyl$^{(R)}$, which was a mixture of tin powder and tin(II) oxide in tablet form, had,

Table I ^{99m}Tc-Sn radiopharmaceuticals (Blunden et al 1985)

Complex[a]	Tin(II) salt in formulation	Uses
^{99m}Tc/Sn-EHIDA	$SnCl_2 \cdot 2H_2O$	Hepatobiliary scintigraphy
^{99m}Tc/Sn colloid	SnF_2	Liver and spleen scintigraphy
^{99m}Tc/Sn-gluconate	$SnCl_2 \cdot 2H_2O$	Kidney and brain scintigraphy
^{99m}Tc/Sn-DTPA	$SnCl_2 \cdot 2H_2O$	Kidney and brain scintigraphy
^{99m}Tc/Sn-DMSA	$SnCl_2 \cdot 2H_2O$	Kidney scintigraphy
^{99m}Tc/Sn-tetracycline	$SnCl_2 \cdot 2H_2O$	Heart, kidney and gall bladder scintigraphy
^{99m}Tc/Sn-MMA	$Sn(OH)_2$[b]	Lung scintigraphy
^{99m}Tc/Sn-MDP	SnF_2	Skeletal scintigraphy
^{99m}Tc/Sn-RBC	SnF_2 or $Sn_2P_2O_7$	Red blood cell labelling

[a] EHIDA = 2,6-diethylphenylcarbamoylmethyliminodi(acetic acid); DTPA = diethylenetriaminepentaacetic acid; DMSA = dimercaptosuccinic acid; MAA = Macroaggregated human albumin; MDP = methylene diphosphonic acid; RBC = red blood cells.

[b] Hydrous tin(II) oxide.

until recently, been marketed in the UK as a treatment of various skin complaints such as acne, boils, carbuncles and styes (Sadler 1982).

The above applications represent the sole human pharmacological uses of tin chemicals. The remainder of this article reports on areas where these compounds are showing potential for future utilisation.

TIN PROTOPORPHYRIN FOR THE TREATMENT OF NEONATAL JAUNDICE

Neonatal jaundice is a common and sometimes severe condition, which may produce neurotoxic effects. It arises when the liver of the newborn is not sufficiently developed to enable it to detoxify the bile pigment, bilirubin, and hence the bile pigment accumulates in the blood stream (hyperbilirubinemia) during the first week or more of neonatal life (Drummond and Kappas 1982). Bilirubin arises from the degradation of heme (iron protoporphyrin IX), which occurs by two enzymic reactions, the first of which, involving heme oxygenase is the rate determining step (Rideout et al 1987):

$$\text{heme} \xrightarrow{\text{heme oxygenase}} \text{biliverdin} \xrightarrow{\text{biliverdin reductase}} \text{bilirubin}$$

Since dichloro(protoporphyrin IX)tin(IV) (tin protoporphyrin, Sn-heme, SnPP) [1] was known to be a potent inhibitor of heme oxidase activity in the liver, spleen, kidney and skin of rats, its effect on neonatal jaundice was studied. When Sn-heme was administered to newborn rats an immediate (within 1 day) and significant lowering of serum bilirubin levels occurred and this decline continued throughout the period when the serum concentration of the bile pigment in control neonates was increasing or remaining above normal adult levels (Drummond and Kappas 1981), Figure 1. Furthermore in Sn-heme treated animals near normal levels of serum bilirubin levels were quickly reached (day 3) and were maintained throughout the 42 days of the study, Figure 1. The effectiveness of Sn-heme to rapidly reduce the amount of bilirubin in the blood stream has been demonstrated for a number of other laboratory animals *ie* mice (Sassa et al 1983; 1985), rats (Nagae et al 1988); Sisson et al 1988) and monkeys (Cornelius and Rodgers 1984; Cornelius et al 1985) and also in human adults (Kappas et al 1984; Berglund et al 1988).

Other clinical events, such as congenital anemias, thalassemia and sickle cell anemias, as well as various forms of liver diseases, can also lead to the development of hyperbilirubinemia. In such cases the concentration of bilirubin in the serum of such individuals rarely reaches the high levels

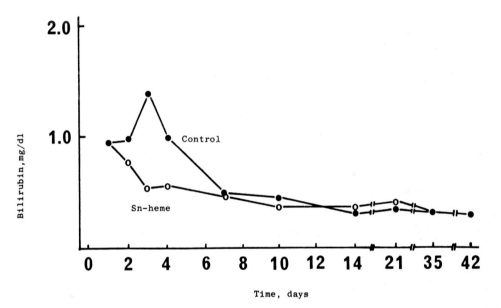

Figure 1. Effects of Sn-heme on total bilirubin levels in the neonate during the first 6 weeks after birth. (Drummond and Kappas 1981)

found in the new born. However it does attain harmful levels and should be controlled. Sn-heme has been found to be effective in the treatment of such conditions (Kappas et al 1984; Rideout et al 1987).

It was clear from the above studies that the administration of Sn-heme to laboratory animals, to normal volunteers and to patients with hepatic dysfunction, prevents or reduces a variety of experimentally induced or naturally occurring forms of jaundice. When used in small single or multiple doses in clinical situations on human subjects the compound was found to be innocuous. The only side effect to be observed was that single doses > 1µmol/kg body weight produced a transient photosensitivity in some individuals. Having thus established the good activity and apparent lack of harmful side effects of Sn-heme, a study of the effect of this compound on new born infants with direct Coombs-positive ABO incompatibility, a group which is highly susceptible to the onset of hyperbilirubinemia, was performed (Kappas et al 1988). A total of 69 control and 53 treated infants were used in two studies: In the first a single dose of 0.5µmol/kg body weight was administered via intramuscular injection, whereas in the second; one, two or three doses of 0.75µmol/kg were used. In both studies incremental changes in plasma bilirubin levels were lower in treated groups than in the controls. (Table 2, Figures 2 and 3). In study 1 the differences did not reach statistically significant levels within 96h, however in Study 2, which used higher doses of Sn-heme, statistically significant differences were observed, these began within the 24h to 48h time period, after administration of the compound, and extended to the end of the study (96h). Figures 2 and 3 clearly show that the Sn-heme treated groups always had lower plasma bilirubin values than the control groups, and that in Study 2 the hyperbilirubinemia of infants treated was diminished earlier than that of the controls, and to a greater degree than in Study 1.

In addition to decreasing hyperbilirubinemia a secondary effect from the use of Sn-heme was a decrease in the number of infants requiring phototherapy. Thus in Study 1 34.2% of the controls and 17.2% of the treated infants received phototherapy,

Table 2. Incremental Changes in Plasma Bilirubin Concentrations Over Initial Bilirubin Values in Control and Sn-Protoporphyrin-Treated Infants* (Kappas et al 1988)

Time After Initial Blood Sample (h)	Study Group 1		Study Group 2	
	Control	Sn-Protoporphyrin Treated	Control	Sn-Protoporphyrin Treated
12-24	2.55 ± 0.29 (37)	1.78 ± 0.24 (29)	2.79 ± 0.28 (24)	2.21 ± 0.26 (22)
24-48	4.55 ± 0.50 (37)	3.71 ± 0.41 (29)	5.14 ± 0.58[a] (23)	3.87 ± 0.38[a] (21)
48-72	5.39 ± 0.6 (40)	4.87 ± 0.60 (29)	6.71 ± 0.62[b] (27)	4.78 ± 0.62[b] (22)
72-96	6.16 ± 0.80 (29)	5.14 ± 0.92 (21)	7.62 ± 0.84[c] (24)	5.16 ± 0.63[c] (21)

*Results are mean milligrams per deciliter = SE. Numbers in parentheses are numbers of infants studied. t test results: [a] $p<.08$. [b] $p<.03$. [c] $p<.03$.

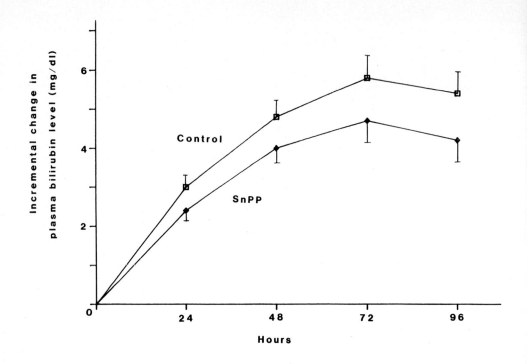

Figure 2. Incremental plasma bilirubin levels over initial levels (mean values = SE) for control and Sn-protoporphyrin (SnPP)-treated infants in study 1. (Kappas et al 1988)

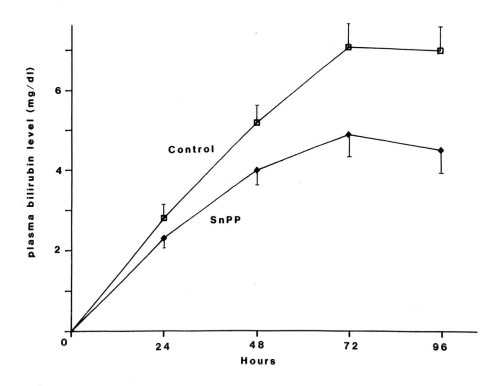

Figure 3. Incremental plasma bilirubin levels over initial levels (mean values = SE) for control and Sn-protoporphyrin (SnPP)-treated infants in study 2. Values for statistical significance of differences observed between two groups were: at 48 hours P<.05: at 72 hours P<.02: at 96 hours P<.02. (Kappas et al 1988)

while for study 2 the figures were; 46.4 and 29.2% respectively. Furthermore there was strong evidence that Sn-heme treatment alone was more effective than phototherapy in moderating postnatal hyperbilirubinemia. Sn-heme also decreased by 42% the number of phototherapy hours in the combined treatment groups as compared with the controls.

Only 2 cases of clinical side effects were noted. These both occurred in Study 2 when cutaneous erythema was observed after phototherapy and both disappeared completely; one during the phototherapy session and the other 72h after its cessation.

This is the first report of clinical trials on human newborns and the results are most encouraging. It is hoped that further studies will confirm the suitability of Sn-heme for such treatments since it would be most advantageous for use in clinical situations where for social or economic reasons other treatments, such as phototherapy, were not available.

An additional problem often encountered in patients with hyperbilirubinemia is the independent accumulation of excess iron. It has been found that Sn-heme greatly enhances the biliary excretion of iron into the intestinal contents, from where the metal is eliminated. In addition by blocking the binding of heme to heme oxygenase the release of iron, which normally occurs during heme catabolism, is prevented. Thus one atom of iron is excreted into the intestine with every molecule of uncatabolised heme (Rideout et al 1987).

Sn-heme can also be used to control the concentrations of tryptophan and serotonin in the brain. Increased concentrations of these chemicals are associated with hepatic encephalopathy and migraine headaches. Such conditions can be alleviated by administration of Sn-heme, since this has the effect of increasing the amount of heme in the liver, which in turn increases the rate of tryptophan metabolism in the liver, thus reducing the amount of tryptophan released into the plasma and finally results in less accumulation of tryptophan and serotonin in the brain. (Rideout et al 1987).

Two further potential uses of Sn-heme have recently been identified: Sn-heme was found to lower heme oxygenase activity in tumour bearing animals to below control levels, this result led to the suggestion that it may be suitable for the protection of normal cells during tumour growth and chemotherapy (Wissel et al 1988). Stout and Becker (1988) have studied the effects of Sn-heme on the hepatic xenobiotic metabolising enzymes in the rat. Their findings suggested that Sn-heme might affect the metabolism of other xenobiotics and this possibility was confirmed by the finding that hexobarbital-induced sleep lasted 4 times longer in Sn-heme treated rats than in controls.

Two other tin metalloporphyrins have been identified as having similar beneficial properties, tin mesoporphyrin(IX)

(SnMP)[2] and tin diiododeuteroporphyrin (SnI$_2$DP)[3]. Sn-heme and SnI$_2$DP tend to exhibit similar activity, while SnMP is the most effective (Rideout et al 1987). Sn-heme and SnMP may both

R^1 = CH$_2$CH$_2$CO$_2$H

{2} R = Et

{3} R = I

produce phototoxic side effects such as skin rashes, flushed skin and general discomfort, which arise after treatment, on exposure to sunlight or fluorescent light, whereas SnI$_2$DP is much less active (Rideout et al 1987; Delaney et al 1988). In contrast, all three were reported to be extremely poor photosensitisers when excited in the spectral region commonly used in phototherapy (Delaney et al 1988). However, as indicated above, when Sn-heme was used to treat human newborns only 2 cases of clinical side effects were recorded, and these arose after phototherapy and they both disappeared completely. Thus these tin porphyrins represent a new, potent and convenient means of treating neonatal jaundice and related conditions.

ANTITUMOUR PROPERTIES OF TIN CHEMICALS

Until recently organic molecules have dominated the chemotherapy of cancer. In 1979 cis-diamminodichloroplatinum(II), cisplatin [4] was approved for use, and is now the leading anticancer drug in the USA (Sykes 1988). Cisplatin exhibits activity against a large number of tumours, being particularly potent in the treatment of testicular tumours, ovarian carcinomas and lung cancers (Cleare 1974; Sadler 1982; Sykes 1988). The success of cisplatin has led not only to the development of second generation platinum complexes such as carboplatin[5], which is approved for use in the UK, iproplatin[6] and spiroplatin[7], all of which are less toxic and more active than cisplatin, but also to the investigation of the antitumour properties of many other metal compounds. (Cleare 1974; Sykes 1988)

As far as tin compounds are concerned, early studies showed that tetraphenyltin, hexaphenylditin, triphenyltin bromide and triphenylpropyltin were inactive against transplanted mouse cancers (Krause 1929). Similarly, tin(II) chloride was found to be inactive against spontaneous tumours in mice and rats (Kanisawa and Schroeder 1967; 1969). However, in 1972, Brown demonstrated that tumour growth in mice was retarded by triphenyltin acetate, but not by the chloride, when administered either in the food or by injection. During the period 1973-1977, the Institute for Organic Chemistry TNO, Utrecht, The Netherlands, carried out research, which included the submission of organotin compounds for screening as antitumour agents, sponsored by the International Tin Research Institute. A wide variety of organotins was tested but none were found to be of sufficient activity to warrant further screening (Crowe 1980). Subsequent work by this Dutch group resulted in the identification of a number of active diorganotin oxides (R_2SnO), hydroxides $R_2Sn(OH)X$, distannoxanes $(XR_2Sn)_2O$ and di(methylcarbonylmethoxides) $R_2Sn(CH_2.CO.Me)_2$ which either contain a tin oxygen bond or a capable of generating such a bond on hydrolysis (Bulten and Budding 1981; Meinema et al 1985).

The various investigations involving cisplatin analogues had shown that activity was usually associated with square planar

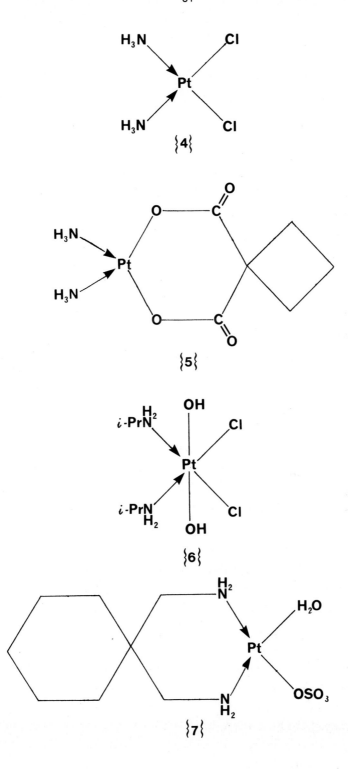

platinum(II)[8] and octahedral platinum(IV)[9] complexes which possessed the following features; two *cis*-nitrogen donor ligands each bearing at least one hydrogen atom, and two good leaving groups eg. Cl^-, Br^-, RCO_2^- etc., also in a *cis* configuration (Cleare 1974; Sadler 1982). Tin compounds do not form square planar complexes, however, diorganotin dihalides (R_2SnX_2) will interact with two monodentate ligands (L) or one bidentate ligand (L_2) to form octahedral complexes ($R_2SnX_2.L_2$) [10-13]. The antitumour activity of some 115 such complexes (R = Me, Et, Pr, Bu, Ph, Bz, Oct; X = F, Cl, Br, I, NCS; L = N- or O- donor ligands) has been studied (Crowe and Smith 1980; Crowe et al 1980, Crowe et al 1984a). The majority of the ligands used were bidentate to ensure that the resulting octahedral complex possessed *cis*-halogens [12-13] which as mentioned previously had been shown to be an essential requirement for activity. The nitrogen donor ligands used contained aromatic nitrogen atoms, since analogous aliphatic amine complexes of tin tend to be hydrolytically unstable towards atmospheric moisture. However it should be noted that dimethyltin dihalide (Me_2SnX_2; X = Cl, Br) adducts with ethylene diamine and with 1, 4-diaminobutane have been screened against P388 lymphocytic leukaemia in mice, but were found to be inactive and quite toxic (Eng and Engle 1987). Many of the $R_2SnX_2.L_2$ complexes exhibited reproducible activity against P388 leukaemia in mice. In most cases the parent diorganotin dihalide and the free ligand were inactive and so it was thought that the activity seen was a function of the complex.

Another class of active antitumour compounds, which are structurally similar to cisplatin, are the metallocene dichlorides[14] (M = transition metal). Of these the titanium and vanadium derivatives were found to be most effective (Köpf and Köpf-Maier 1983; Köpf-Maier et al 1988). The antitumour activity of the metallocene dichlorides as well as that of cisplatin has been related to the magnitude of the ClM̂Cl bond angle, in that compounds for which the ClM̂Cl angle was <95° were active (Köpf and Köpf-Maier 1983). When this theory was examined for the $R_2SnX_2.L_2$ complexes it was found that both active and inactive

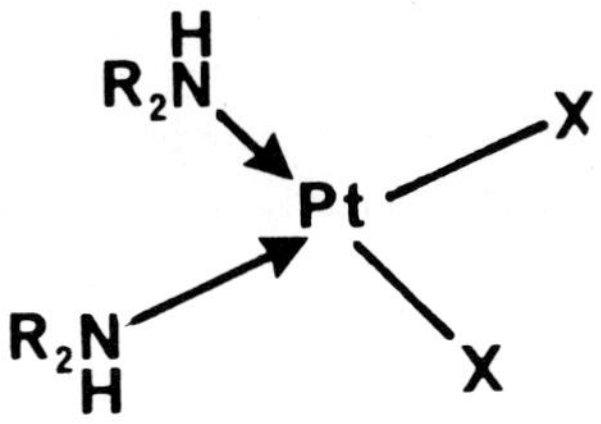

{8}

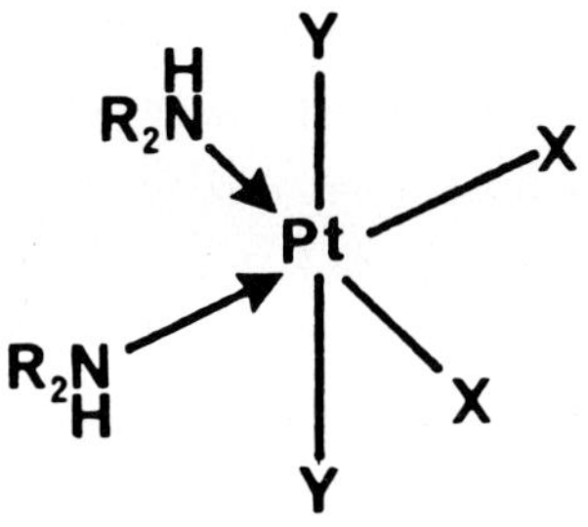

{9}

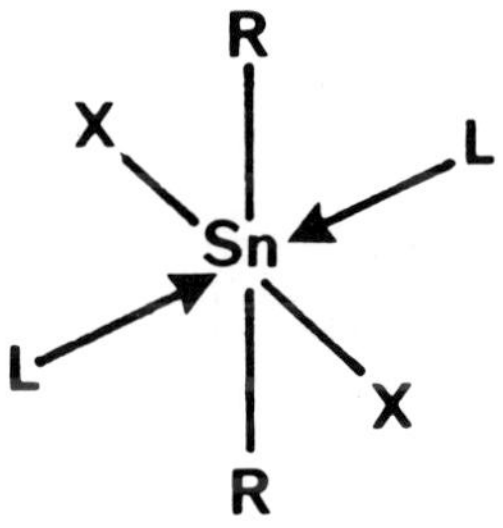

{10}

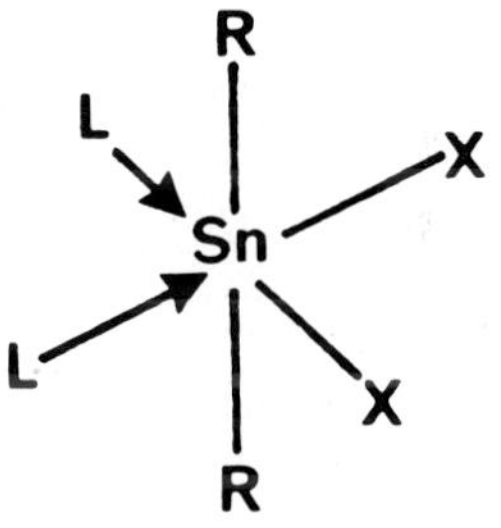

{11}

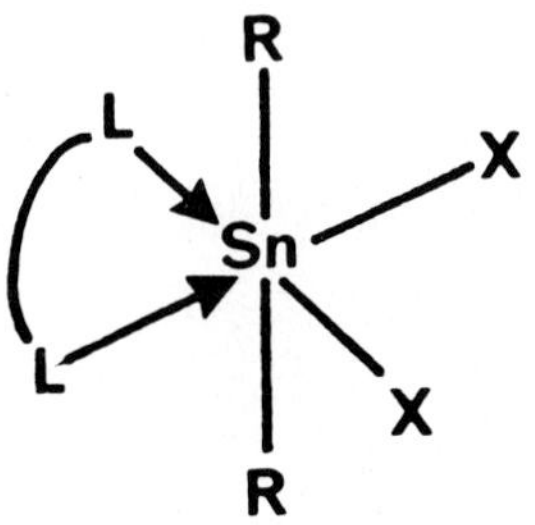

{12}

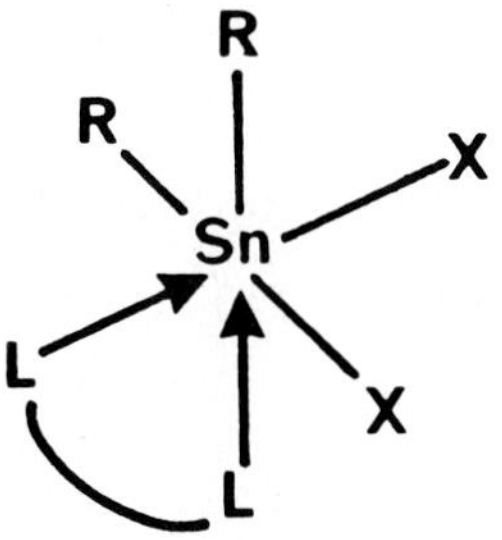

{13}

complexes had ClMCl bond angles of 103-105°, which were above the limiting value proposed by Köpf and Köpf-Maier. But further examination of the structural data revealed that active tin

complexes had average tin-nitrogen bond lengths >2.39Å, whereas
inactive complexes had Sn-N bond lengths <2.39Å. This observation led to the suggestion that the more stable complexes
(short Sn-N bonds) had lower activities, which in turn implied
that a predissociation of the bidentate ligand may be an
important step in the mode of action of the tin complexes
(Crowe et al 1984b, Crowe 1988a).

{14}

{15}

Recently two air stable, main group analogues of the metallocene dichlorides, decaphenyl- stannocene and -germanocene
$[\eta^5-(C_6H_5)_5C_5]M^{II}$ (M = Sn, Ge)[15], have been screened against
Ehrlich ascites tumour in mice. Both derivatives gave cure
rates of 40-90% dependent on the dose administered, but the
tin derivative caused toxic deaths at doses > 440mg/kg, while

for the germanocene no toxic deaths occurred at doses < 700mg/ kg. (Köpf-Maier et al 1988). The activity of these compounds was surprising since no covalently bound acido ligands, which were structural features of active platinum, titanium, vanadium and tin derivatives, were present. In addition previously studied substituted cyclopentadienyls were invariably inactive.

Many other diorganotin compounds containing Sn-O, Sn-N or Sn-S bonds have also been screened against lymphocytic leukaemia in mice (Barbieri et al 1982; Haiduc et al 1983; Saxena and Tandon 1983; Gielen et al 1984; Huber et al 1985; Crowe 1987; Crowe 1988a). An examination of the screening data for these compounds as well as the $R_2SnX_2 \cdot L_2$ complexes reveals that the diethyl- and/or diphenyl-tin derivatives when active usually exhibited the highest activity, Figure 4. However, this activity was considerably lower than that of the platinum complexes against the same tumour system.

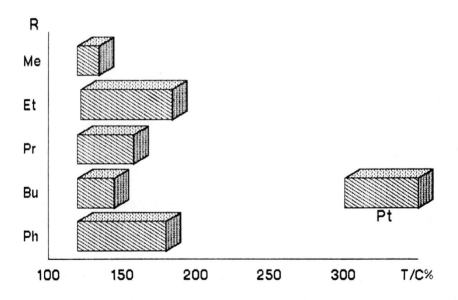

Figure 4.   Summary of the Activity of R_2SnX_2 (X = halogen, O,N,S) against P388 Leukaemia in mice (Crowe 1988b).

Tin compounds have also shown <u>in vivo</u> activity against other murine tumour systems; $Et_2SnCl_2 \cdot PBI$ and $Ph_2SnCl_2 \cdot TMphen$ were active against the renal adenocarcinoma tumour system (Atassi 1985); Bu_2SnCl_2 was active against Ehrlich ascites and IMC carcinoma (Wada and Manabe 1986); while dihalobis(benzoylacetonato)tin(IV) compounds, $X_2Sn(bzac)_2$, have been found to be active against sarcoma 180 tumour in mice (Keller et al 1982). This is the first report of good activity by a tin compound against a solid tumour and the T/C value of 230% for the dichloride is very encouraging.

Other studies using dibutyltin dichloride have shown that it will inhibit the onset of induced cancers. Thus in a two-stage mouse-skin carcinogenesis system of initiation and promotion, the optimal dose of dibutyltin inhibited more strongly the promotion stage (Arakawa 1986). While the intragastric application of dibutyltin dichloride (DBTC) after a single dose of N-nitrosobis(2-oxopropyl)amine (BOP) administration significantly reduced the induction of ductal adenocarcinomas in hamsters, whereas when the compound was given before carcinogen treatment the incidence of pancreatic cancer was unaffected (Takahashi et al 1983). In contrast, when DBTC was given before the administration of BOP, once a week for five weeks, a significant inhibitory effect on pancreatic carcinoma induction was seen, whereas when DBTC was given after the last BOP treatment no such influence was seen (Jang et al 1986). The results of these two studies appear to be contradictory, in both cases DBTC exhibits inhibitory action on BOP induced pancreatic carcinogenesis, but at differing times of administration. However, Jang et al (1986) have clarified this by pointing out that the two investigations differed with regard to the duration of carcinogen exposure. Thus prior DBTC insult brought about a reduction in tumour yield only when BOP injections were given over a 5 week period, this suggested the involvement of some relatively long-term influence. Whereas, inhibition due to subsequent DBTC administration was limited to the single BOP injection case.

Cardarelli et al (1984a;b) have investigated the effect of administering tin compounds in drinking water to cancer prone mice. $Bu_2SnCl_2 \cdot bipy$, $Bu_2SnCl_2 \cdot phen$, $Bu_2Sn(histidine)$ or Bu_3SnF when given in this way produced significant reductions in tumour growth rates. These and related studies have led to the suggestion that the tin intake, whatever its form, is biochemically converted to anticarcinogenic organotin entities in the thymus and these are subsequently distributed through the body via the lymphatic system (Cardarelli et al 1983; 1984a; 1984b). Further studies have indicated that the thymic tin compounds are likely to be derivatives of steroids, and it has been shown that organotin steroids, in particular cholesterol derivatives will retard the growth of and even kill malignant tumours or prevent cancerous proliferation (Cardarelli and Kanakkanatt 1985; Cardarelli 1985; Cardarelli and Kanakkanatt 1987; Sherman et al 1987)

Certain tin derivatives have been examined for activity <u>in vitro</u>, Table 3. The dibutyltin derivative of pyridoxime

Table 3.   <u>In vitro</u> Antitumour Studies

Compound	Tumour system	Reference
$Bu_2Sn(pyridoxime)$[16]	L1210 leukaemia	a
$Bu_2Sn(cortexolone)$[17]	P388	"
$Bu_2Sn(erythromycin)$	P815	"
$[Et_2NC(S)S-]_2SnCl_2$	B16 melanoma	b
	3T3 fibroblasts	
$R_2Sn(2,6-pyridinecarboxylate)$[19]	L1210 leukaemia	c
	P388	"
R = n-Bu, t-Bu, Ph	P815	"
	B16 melanoma	
	Lewis lung carcinoma	

a) Gielen et al 1988; b) Carra et al 1988; c) Gielen et al 1987

(vitamin B6) [16] exhibited <u>in vitro</u> activity which was almost ten times greater than that of the clinically used methotrexate, whereas the dibutyltin derivative of cortexolone (11-desoxy-

{16}

{17}

{18}

17α-hydroxycorticosterone)[17] was less active, and the dibutyltin derivative of erythromycin, the structure of which is currently unknown, only shows low activity. Both cortexolone and erythromycin[18] were inactive against all three tumour cell systems (Gielen et al 1988).

Dichlorodiethyldithiocarbomato tin(IV), $[Et_2NC(S)S-]_2SnCl_2$, showed high cytotoxicity (ID_{50} values after exposure for 24 and 48h were 5.10^{-5}M and 3.10^{-5}M on B16-F10 cells and 4.10^{-5}M and 9.10^{-6}M on 3T3 cells, respectively) but low toxicity in vivo (acute LD_{50} i.p. in mice >3000mg/kg). In view of this data it was proposed that the in vivo antitumour activity of this compound would be studied (Carrara et al 1988).

Three diorganotin derivatives of 2,6-pyridinedicarboxylic acid [19] were found to be more active in vitro against L1210, P388 and P815 leukaemias, B16 melanoma and Lewis lung carcinoma than cisplatin[4] (Gielen et al 1987). However when one of these, the di-n-butyl derivative, was screened in vivo against P388 leukaemia in mice, it gave a T/C value of 135% (dose 15mg/kg)which is a typical result for a dibutyltin compound against this tumour system (Gielen et al 1987). In contrast, cisplatin has an activity of T/C = 289% (dose 2mg/kg). The disappointingly low activity of the dibutyltin derivative no doubt arises due to the great difference in conditions between the in vitro and in vivo test systems. The in vitro results suggest that diorganotins have the potential for high activity, while the in vivo results show that more work is necessary to determine how best to deliver them to the site of action.

The majority of the tin compounds which have been screened for antitumour activity were poorly soluble in water, and it has been suggested that this factor is a major obstacle to an improvement in their activity (Atassi 1985). In an attempt to impart water solubility on a diorganotin moiety, anions of the type $[R_2Sn(-SCH_2CH_2SO_3)_2]^{2-}$ were synthesised. Three of these {R = Me, cation = Na^+; R = Et, cation = Na^+, $[C(NH_2)_3^+]$} were administered in solution in saline, while two further covalent derivatives $[Bu_2Sn(penicillamine)_2$ and $Me_2Sn(\underline{N}-benzoylglycine)_2]$ were given in solution in the aqueous medium klucel. Of these five compounds, only $Me_2Sn(N-benzoylglycine)$ was inactive,

{19}

however the activity of the other four was still rather low (P388 leukaemia in mice T/C <140%) (Huber et al 1985). Three further water-soluble organotin molecules [20-22] were found to be inactive against P388 leukaemia in mice, while [22] was also inactive <u>in vitro</u> against P388, L1210 and P815 leukaemia tumour cells (Gielen et al 1988).

{20}   {21}

{22}

Other ways of improving the delivery of diorganotins have been investigated: Perfluorochemicals are considered to be suitable vehicles for the transport of drugs to tumours, and for this reason, a series of compounds, R_2SnX_2 [where R = $CF_3(CF_2)_5CH_2CH_2$; X = Cl, Br] and their adducts with bipy and phen were synthesised and subsequently screened against P388 leukaemia in mice, but all were inactive (De Clerq et al 1984). Similarly inclusion complexes of diorganotin dihalides in β-cyclodextrin were also inactive against the same tumour system (Gielen et al 1988).

MODE OF ACTION

The mode of antitumour action of cisplatin [4] appears to be fairly well established; the complex is believed to lose its chloride ligands and the metal subsequently coordinates with suitably orientated nitrogenous bases on DNA (Prestayko et al 1980). Experimental evidence has indicated that the N-7 nitrogen atoms of two adjacent guanines are the most likely binding sites and this has been supported by the elucidation of the X-ray crystal structure of the covalent adduct of cisplatin with a DNA fragment: <u>cis</u> [$Pt(NH_3)_2d(pGpG)$] (Sherman et al 1985). The d(pGpG) segment, which contains two deoxyguanosine units, is believed to be cisplatin's main target in cancer cells. The X-ray structure clearly showed that cisplatin crosslinked the DNA and that the binding sites were the N-7 nitrogen atoms on two adjacent guanine rings of the same chain. Furthermore, to accommodate the platinum atom in this structure, the guanine molecules were tilted away from their normal stacked positions, thus disrupting the DNA helix and so interfering with replication.

Examination of the data for the various tin derivatives reveals that many more diorganotin compounds exhibit antitumour activity than the corresponding mono-, tri- and tetra-organotins, or the inorganic tin chemicals. While within the diorganotin class the highest activity is given by the diethyl- and diphenyltin derivatives, Table 4. This would suggest that the activity of the diorganotins (R_2SnX_2) in general is controlled by the nature of the R_2Sn moiety. This is not really surprising since the Sn-X (X = halogen, O,N,S) bonds present in these compounds are susceptible to hydrolysis and so all of the diorganotin derivatives would ultimately yield analogous hydrolysed R_2Sn species, which would then be responsible for the antitumour activity observed. Ruisi et al (1985) have demonstrated that

Table 4 R_2Sn compounds with T/C% > 170 against P388 leukaemia (Crowe 1987)

Compound	Dose (mg/kg)	T/C(%)
$Et_2SnCl_2 \cdot PBI$	100	171
$Et_2SnCl_2 \cdot PBI$	40	218[a]
$Et_2SnCl_2 \cdot phen$	25	177
$Et_2SnBr_2 \cdot PBI$	12.5	175
$Et_2SnBr_2 \cdot phen$	25	176
$Et_2SnI_2 \cdot phen$	200	184
$Et_2Sn(NCS)_2 \cdot bipy$	12.5	179
$Et_2Sn[SCH_2CH(NH_2)CO.O]$	25	180
$Et_2Sn(CH_2CO.OMe)_2$	12.5	170
$EtPhSn(CH_2CO.OMe)_2$	50	181
Ph_2SnF_2	-	196
$Ph_2SnCl_2 \cdot 2py$	-	180
$Ph_2SnCl_2 \cdot TMphen$	12.5	173[a]
$Ph_2SnCl_2 \cdot TMphen$	6.25	177
$Ph_2SnCl(OH)$	25	198
$Ph_2Sn[SCH_2CH(NH_2)CO.O]$	50	181
Bu_2SnCl_2	3.0	186[b]

[a] Renal Carcinoma i.p.

[b] Ehrlich ascites

for diorganotin glycylglycinate complexes the observed antitumour activity is not attributable to the coordinated ligand GlyGly^{2-} or to its configuration when chelating to a metal centre, but that the coordinated ligand favours in some way the transport of the drugs into cells and that the antitumour activity arises from $R_2Sn(IV)$ moieties released by slow hydrolysis of the complexes. Thus for the diorganotin compounds, R_2SnX_2 (X = halogen, O,S,N) it would appear that the function of the X_2 group(s) is to aid transport of the active R_2Sn moiety to the site of action, where it is released by hydrolysis. This mechanism helps to explain the variations in activity seen for a series of diorganotin derivatives in which R is kept constant and X is varied. If the compound is too hydrolytically unstable, the R_2Sn species will be released too soon, and if it is too stable it may be released too late or too slowly for activity to be seen. Since the R_2Sn species is of prime importance with respect to antitumour activity, future research should concentrate on exploring the effect of more complex organic groups, R (eg unsaturated, functionally substituted etc), attached to tin. Sherman and Huber (1988) have suggested that the best prospects for antitumour properties in organotin compounds lies in those containing aryl or cycloalkyl groups or possibly biologically active groups, since low molecular weight organotin moieties, R_2Sn and R_3Sn (R = Me, Et, Pr, Bu, Oct) produce adverse effects upon the immune and nervous systems of rodents.

In the case of triorganotin compounds, R_3SnX, low oral doses of tributyl- and triphenyl-tin derivatives when administered over a long period of time produce activity (Brown 1972; Cardarelli et al 1984a;), whereas when they are injected they are usually toxic at high doses and with low doses no activity is seen (Crowe 1980; Huber 1985). Triorganotins are known to undergo dealkylation or dephenylation in vivo to yield the corresponding diorganotin (Wada and Manabe 1986). The half life for tributyltin fluoride to undergo debutylation has been shown to be between 3.7 and 6.6 days (Iwai et al 1979). Thus, during the feeding studies with triorganotin derivatives, loss of one of the organic groups attached to tin would release an R_2Sn species, which could then behave in a similar manner to that previously described for the injected diorganotin derivatives.

The ultimate fate of the active R_2Sn moiety is not known, it may be that it crosslinks DNA in a manner similar to that of cisplatin, certainly diorganotin derivatives of nitrogenous bases e.g. $R_2Sn(adenine)_2$, are known (Barbieri et al 1985). In addition, 5-Fluorouracil (FU) which is structurally similar to nucleic acid bases, on reaction with Et_2SnCl_2.phen, displaced a chloride ligand to yield $Et_2SnCl(FU)$.phen [23] (Jiazhu et al 1988). It is of interest to note that it was claimed that [23] showed higher antitumour activity than either of its individual components, but no data was supplied to support this.

{23}

Thus in summary, many organotins show reproducible antitumour activity in mice and rats, without the toxic side effects associated with cisplatin, however this activity is lower than that of cisplatin. A number of probable factors relating to the mode of action of the tin compounds have been identified;

i) the organic group, R, determines the potential activity;
ii) the X group controls delivery of the active R_2Sn species;
iii) the hydrolytic stability of the Sn-X bonds determine whether the potential activity of the R_2Sn moiety is realised;
iv) the tin atom may interact with DNA in a similar manner

to that of platinum;

but more work is necessary to validate the above statements. Such studies would hopefully lead to a better understanding of the mode of action of the tin compounds, and more importantly to compounds with higher activities.

THE USE OF TIN DERIVATIVES IN THE PHOTODYNAMIC THERAPY OF CANCER

Photodynamic therapy (PDT), involving the use of photosensitisers and red light, is currently being extensively examined as a treatment for solid malignancies in humans. The main advantages of PDT over conventional cancer treatments are its selectivity and low systemic toxicity. In addition PDT may be used to treat regions which have already received maximal doses of conventional radiotherapy. (Morgan et al 1987 b,c; Pottier 1988; van Lier et al 1988).
 Photodynamic action is the term used to describe oxygen dependent photosensitisation and thus distinguish this phenomenon from the photosensitisation that occurs on photographic plates. The term phototoxicity may also be used to describe this phenomenon. Exposure to visible or uv light provides a simple and convenient way to excite atoms or molecules to a reactive state. However many biologically active molecules do not photoexcite directly since they do not absorb electromagetic radiation of these wavelengths. It therefore becomes necessary to use a sensitiser molecule, which will absorb visible or uv light and then in turn transfer this energy to the molecule of interest, thus providing a convenient means of indirect photosensitisation (Pottier 1988). Red light is preferred in such treatments since at these wavelengths tissue penetration is maximised (van Lier et al 1988). Photodynamic therapy makes use of photosensitising agents, such as certain porphyrin derivatives, which selectively accumulate in tumour tissue, and which on exposure to light induce, most likely via singlet oxygen formation, either plasma membrane or intracellular damage, which leads in turn to tumour necrosis (Pottier 1988). In addition,

tumour blood flow is reduced and it is believed that the resulting anoxia contributes to the cell death which occurs (Morgan et al 1987c).

The most widely used photosensitisers for clinical trials of PDT have been haematoporphyrin derivative (HPD) and its putative active components the dihaematoporphyrin ethers and/ or esters (DHE). These agents appear to be effective, but have the disadvantages of being mixtures of various porphyrin species, each of whose contribution to the total biological effect is unclear, and their absorption maxima (max 630nm) are poor in the red region of the visible spectrum, a region in which tissue penetration is optimum. For these reasons the effects of other photosensitisers, which absorb more strongly in the red region of the electromagnetic spectrum than HPD and DHE, are currently being examined, e.g. phthalocyanines, chlorins, purpurins, tetra(hydroxyphenyl)-porphyrins and verdins (Morgan et al 1987 a,b,c; Moreno et al 1988). A further development of these studies has been the examination of the effects of the metalloderivatives of some of these agents.

Magnesium, chloroaluminium, dichlorotin, copper, zinc, zirconium, fluorochromium, iron, cobalt nickel and palladium derivatives of phthalocyanines [24] have been examined for efficacy as PDT agents (Ben-Hur and Rosenthal 1985 a,b; Reddi 1986; Chan et al 1987). Three of these metallophthalocyanine dyes, chloroaluminium, chloroaluminium sulphonated and dichlorotin, exhibited differential phototoxicity to produce total _in vitro_ cell death after red-light irradiation, but little or no cytotoxic effects after exposure to room light. These properties have led to the suggestion that these three dyes may prove to be useful agents in PDT and suggested that _in vivo_ studies should be performed (Chan et al 1987).

Morgan et al (1987a,c, 1988) have studied the photodynamic activity of some purpurins and metallopurpurins, Figure 5, when combined with red light, in the treatment of transplantable, FANFT induced, bladder tumours in rats. Table 5 summarises the histological results of the three most effective purpurins and the three most effective metallo-derivatives. It can be clearly seen that the two tin derivatives had the greatest

Table 5. Histology of Purpurins and Metallopurpurins (Morgan et al 1987c)

Drug	4 Hour Treatment	24 Hour Treatment
NT1	Vascular stasis, haemorrhage, vacuolization, some viable cells.	Extensive necrosis, fewer viable cells.
NT2	Vascular stasis, haemorrhage, vacuolization, some viable cells.	Extensive necrosis, fewer viable cells.
ET2	Vascular stasis, haemorrhage, vacuolization, some viable cells.	Extensive necrosis, fewer viable cells.
ZnET2	Haemorrhage, cytoplasmic vacuolizations, vascular stasis, extensive necrosis.	Extensive haemorrhage, extensive tumor necrosis.
SnET2	Extensive haemorrhage, cytoplasmic vacuolization, vascular stasis.	Complete necrosis.
ZnNT2H2	Haemorrhage, cytoplasmic vacuolization, vascular stasis.	Extensive necrosis.
SnNT2H2	Extensive haemorrhage, cytoplasmic vacuolization, vascular stasis.	Complete necrosis.
AgNT2H2	Minimal haemorrhage, mostly viable cells.	Some haemorrhagic necrosis.
ZnNT1	Haemorrhage, cytoplasmic vacuolization, vascular stasis.	Extensive necrosis.

Table 6. Dose Response Analysis for Free Base Purpurins and Metallopurpurins (Morgan et al 1987c)

		Dose (mg/kg)							
		5.00		2.50		1.00		0.50	
		Control	Treated	Control	Treated	Control	Treated	Control	Treated
NT1	mean (mg) ± SE			1.307 0.239	0.231 0.152	1.314 0.427	0.216 0.107		
	p value			<0.01		<0.01			
	cure rate			40%		20%			
NT2	mean (mg) ± SE	0.155 0.062	0.000 0.000	0.163 0.024	0.018 0.012	0.154 0.052	0.054 0.013		
	p value	<0.01		<0.02		NS			
	cure rate	100%		60%		0%			
ET2	mean (mg) ± SE			0.375 0.084	0.000 0.000	0.359 0.081	0.061 0.027		
	p value			<0.02		<0.02			
	cure rate			100%		40%			
ZnET2	mean (mg) ± SE			0.355 0.109	0.011 0.006	0.258 0.073	0.061 0.029	0.345 0.046	0.108 0.046
	p value			<0.05		<0.05		<0.02	
	cure rate			30%		20%		0%	
SnET2	mean (mg) ± SE			0.144 0.046	0.000 0.000	0.181 0.058	0.008 0.008	0.256 0.112	0.050 0.026
	p value			<0.05		<0.05		NS	
	cure rate			100%		80%		0%	
SnNT2HT2	mean (mg) ± SE			0.860 0.290	0.000 0.000	0.116 0.025	0.000 0.000	0.064 0.009	0.000 0.000
	p value			<0.05		<0.02		<0.005	
	cure rate			100%		100%		100%	

{24}   R = H, SO₂OH

effect. These six agents were then subjected to a dose response analysis, Table 6 ET2, SnET2 and SnNT2H2 all produced 100% cure rates at a dose of 2.50 mg/kg and the latter compound exhibited this cure rate at the lower doses of 1.00 and 0.50 mg/kg. It should be noted that when SnET2 was administered at doses of 30-fold the therapeutic dose no ill effects were seen. In addition SnET2 was found to reduce tumour blood flow, this phenomenon has also been observed for porphyrin and phthalocyanine based PDT and it has been suggested that this disruption in tumour blood flow produces anoxia which is responsible for the cell death observed.

The metallopurpurin SnET2 [25], a metalloverdin ZnSn1 [26] and a chlorin P^3 [27], were used in a study to determine the long term effects of these potent photosensitisers on FANFT induced bladder tumours in rats (Morgan et al 1987b). Histological examination, 4h after treatment with 1mg/kg of SnET2 administered via an emulsion or a liposome, revealed that extensive haemorrhage, oedema, cell vacuolization and necrosis had occurred within the tumours. The other two agents under the same regimes produced only partial necrosis and minor vascular effects, with many viable cells. Tumour regrowth was measured 12 days after treatment, Figure 6. SnET2 had a significant effect on the growth kinetics of tumours at both 1.0

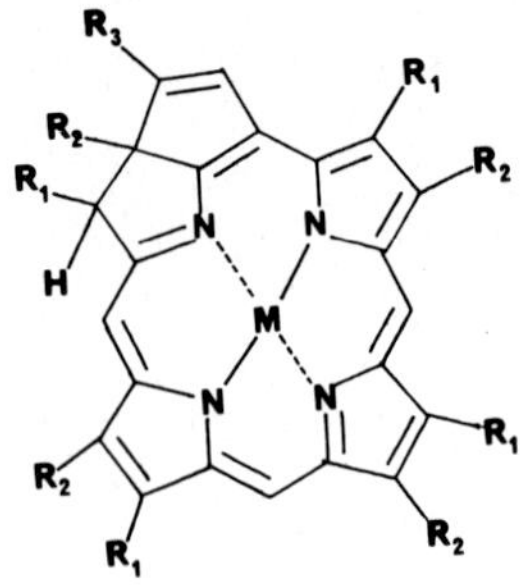

Figure 5. Purpurins and metallopurpurins studied for photo-dynamic activity (Morgan et al 1987c).

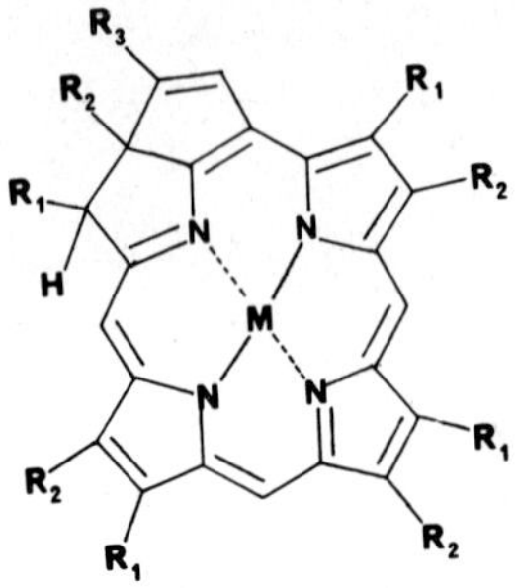

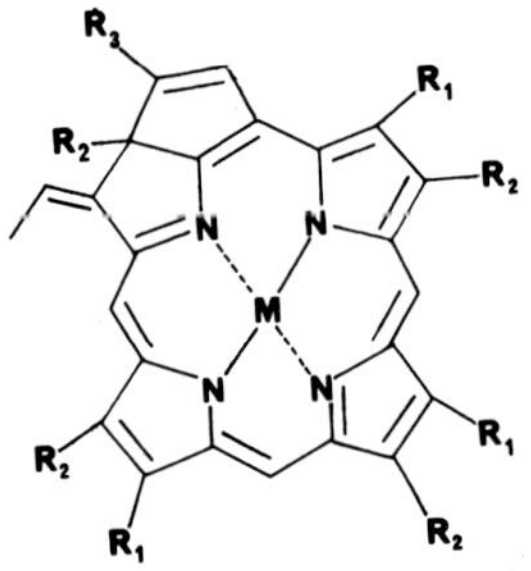

Drug	Fig.	R_1	R_2	R_3	M
NT1	C	CH_2CH_3	CH_2CH_3	$CO_2CH_2CH_3$	2H
NT2	A	CH_2CH_3	CH_2CH_3	$CO_2CH_2CH_3$	2H
NT2H2	B	CH_2CH_3	CH_2CH_3	$CO_2CH_2CH_3$	2H
ET2	A	CH_3	CH_2CH_3	$CO_2CH_2CH_3$	2H
ZnET2	A	CH_3	CH_2CH_3	$CO_2CH_2CH_3$	Zn
SnET2	A	CH_3	CH_2CH_3	$CO_2CH_2CH_3$	$SnCl_2$
ZnNT2H2	B	CH_2CH_3	CH_2CH_3	$CO_2CH_2CH_3$	Zn
AgNT2H2	B	CH_2CH_3	CH_2CH_3	$CO_2CH_2CH_3$	Ag
ZnNT1	C	CH_2CH_3	CH_2CH_3	$CO_2CH_2CH_3$	Zn

and 0.5mg/kg either in emulsion or liposome. At the higher dose level 100% cures were given irrespective of the vehicle. While at the lower dose 20% of the emulsion treated animals

were tumour free, whereas no liposome treated animals were tumour free. In cases where viable tumours remained, both emulsion delivered and lipsome delivered SnET2 were effective in reducing tumour volume (40 and 58% respectively). No comparable cures were observed with either the verdin ZnSN1 or the chlorin P^3 and their effects on tumour reduction were less than those for the tin derivative.

{25}

{26}

{27}

These studies thus demonstrate that low doses of dichlorotinpurpurins are capable of causing extensive tumour necrosis when combined with red light and that these agents are more potent than both the parent purpurin and an equivalent concentration of the widely studied photosensitiser HPD. Another advantage of these metallo-derivatives is that they have large absorption peaks in the red region of the visible spectrum which is an area with excellent tissue penetrating properties, whereas HPD and DHE only have weak absorptions in this region. In addition

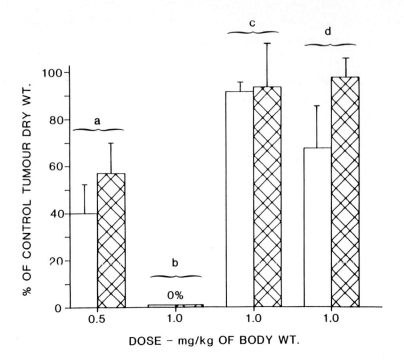

Figure 6. Tumour dry weight of emulsion (open box) and liposome (patterned box) tumours; a and b = SnET2; c = ZnSN1; d = P^3 (Morgan et al 1987b)

these metalloderivatives are pure compounds, unlike HPD and DHE which are ill-defined mixtures. Furthermore the in vitro results involving dichlorotin phthalocyanines suggest that another class of tin containing PDT agents may have been identified. Subsequent studies are clearly needed to further investigate the activity and toxicity of these tin derivatives, however the studies described above suggest that these tin compounds have a promising future as PDT agents for the treatment of solid tumours.

TIN DERIVATIVES AS ANTIVIRAL AGENTS

Viral diseases in man have been estimated to be responsible

for more than 60% of the illnesses that occur in the developed countries of the world. Such infections include the common cold, bronchitis, hepatitis, rabies, polio myelitis, gastroenteritis, influenza, chicken pox, measles, mumps, herpes and AIDS. In addition, viruses have been implicated in a variety of other diseases ranging from rheumatoid arthritis, diabetes, multiple sclerosis, and cervical cancer to congenital heart disease, atherosclerosis, and many other chronic and degenerative processes (Robins 1986).

A series of diorganotin dihalide complexes, $R_2SnX_2 \cdot L_2$, which were modelled on the antitumour agent, cisplatin, showed antitumour activity. Since cisplatin had also been shown to possess antiviral activity, the antiviral activity of the tin complexes was investigated, Table 7 (Ward et al 1988; 1989). The $R_2SnX_2 \cdot L_2$ complexes exhibited weak in vitro antiviral activity against certain DNA viruses; they exhibited selectivity indexes SI = CD_{50}/ED_{50}) in the range of 1 to 10 against three different strains of herpes simplex virus [HSV-1(F), HSV-1(KOS) HSV-2(G)]; while only $Me_2SnBr_2 \cdot$phen demonstrated activity (SI >10) against vaccinia virus. To put these results into perspective, acyclovir, a widely used antiherpes agent has an SI = 5000 against both HSV-1 and HSV-2, while for vaccinia, the nucleoside analogues neplanocin A has an SI = 1300, and 3-deazaaristeromycin has an SI >570.

With a few exceptions the organotin complexes showed no in vitro activity towards a number of RNA viruses. Thus none of them were effective inhibitors of vesicular stomatitis Coxsackie type B4, Sindbis, Semlikiforest or parainfluenza type 3 viruses. Only $Et_2SnBr_2 \cdot$phen and $Ph_2SnBr_2 \cdot$phen showed marginal inhibition against Sindbis and Semliki forest virus respectively. Modest selectivity indexes (>10) were observed for $Me_2SnBr_2 \cdot$PBI and $Me_2SnBr_2 \cdot$phen against Coxsackie virus type B-4, but the selectivity indexes were far surpassed by both neplanocin A (SI = 250) and 3-deazaaristeromycin (SI>200).

Because of the considerable urgency to synthesise and develop agents that can prove efficacious in the treatment of AIDS, the organotin complexes were investigated for their ability to inhibit HIV-1 reverse transcriptase. Unfortunately none of the complexes were found to be inhibitory at the highest concentration ($200\mu g/cm^3$) used. Furthermore, when these complexes were examined for their inhibitory effect on the cytopathogenicity of HIV-1 in human T-lymphocyte MT4 cells, none of them were active at subtoxic concentrations.

In view of the relatively low values of the SIs it was suggested that this particular group of antitumour active organotin complexes should no longer be considered as viable antiviral agents.

Other organotin derivatives are currently being explored (Ward, Taylor, Crowe, unpublished data).

OTHER PHARMACEUTICAL USES

Some organotins have been identified as having activity against certain parasitic diseases. The tropical disease, Leishmaniasis, is a parasitic infection of the skin or viscera in humans which is transmitted by small blood-sucking sand flies. It was found that dioctyltin maleate, at a dose level of 200mg/kg was amongst the most active compounds tested against *Leishmania major* in mice, and that it also displayed activity albeit to a lesser extent against *L. mexicana amazonensis* and *L. infantum* (Peters et al 1980). Recent work by the International Tin Research Institute has confirmed this activity and *in vivo* studies are currently being performed (Crowe unpublished data). A series of organotin(IV) complexes of Schiff bases containing sulphur and fluorine have been tested *in vitro* for amoebicidal activity against *Entameba histolytica* (Saxena et al 1982). Some of these compounds were found to be active at low doses, while one complex, tributyltin (2-fluorobenzaldehyde-*S*-benzyl dithiocarbazate)[28] showed activity greater than the important and currently used drug, emetine.

The antiinflammatory effects of Bu_2SnCl_2 and Ph_3SnCl have been investigated using the carrageenan edema assay (Arakawa and Wada, 1984). Both compounds produced dose-dependent inhibition of edema formation. The inhibition produced by Bu_2SnCl_2 was very similar to that produced by an equal dose of

{28}

hydrocortisone.

Tin(II) chloride has been shown to prevent the development of hypertension in spontaneously hypertensive rats, whereas when administered to normal rats, no effect on blood pressure was seen (Sacerdoti et al 1989). This effect is thought to arise since Sn^{2+}, which is a potent inducer of renal heme oxygenase, caused a depletion of renal cytochrome P-450, which in turn produced a reduction in the formation of cytochrome P-450-dependent metabolites of arachidonic acid, the presence of which is believed to be responsible for the elevation of blood pressure observed in these young rats. Normal blood pressure was maintained up to at least seven weeks beyond the discontinuance of $SnCl_2$ treatment.

Various dialkyltin compounds exhibit anthelmintic properties and a number of commercial preparations for veterinary use, particularly for poultry are available (Blunden et al 1985). However, their suitability for human use does not appear to have been explored.

CONCLUSION

A number of areas in which tin chemicals exhibit therapeutic activity have been identified, of these the use of tin-heme to control neonatal jaundice, and the use of the structurally similar tin-purpurins in the photodynamic therapy of cancer, would appear to be the most likely areas from which a new tin

Table 7. Antiviral selectivity indexes for $R_2SnX_2L_2$ derivatives. (Ward et al 1989)

Selectivity indexes (S.I.)[a]

Compound	HSV-1(KOS)	TK-HSV-1 (B2006)	HSV-2(G)	Vaccinia virus	Vesicular stomatitis virus
$(CH_3)_2SnBr_2 \cdot PBI$	>1	>1	>1	>1	>1
$(CH_3)_2SnBr_2 \cdot phen$	>10	>10	>10	>10	>1
$(C_2H_5)_2SnCl_2 \cdot PBI$	>10	>1	>1	>1	>1
$(C_2H_5)_2SnCl_2 \cdot phen$	>10	>10	>4	>1	>1
$(C_2H_5)_2SnBr_2 \cdot PBI$	>4	10	>1	>1	>1
$(C_2H_5)_2SnBr_2 \cdot phen$	4	20	20	1	1
$(C_6H_5)_2SnCl_2 \cdot PBI$	>1	>1	>2.5	>1	>1
$(C_6H_5)_2SnCl_2 \cdot phen$	>1	>1	>1	>1	>1
$(C_6H_5)_2SnBr_2 \cdot PBI$	2	1	2	1	1
$(C_6H_5)_2SnBr_2 \cdot phen$	5	<2.5	5	1	1
$(C_6H_5)_2SnCl_2 \cdot 2DMSO$	>1	>1	>1	>1	>1

[a] S.I. = ratio of CD_{50} to ED_{50}.

Table 7 continued overleaf ....

Table 7 continued ....

Compound	Selectivity indexes (S.I.)[a]			
	Coxsackie virus type B4	Sindbis virus	Semliki forest virus	Parainfluenza virus type 3	HIV-1
$(CH_3)_2SnBr_2 \cdot PBI$	>10	1	<1	<1	<1.5
$(CH_3)_2SnBr_2 \cdot phen$	>10	>4	<1	<1	<2
$(C_2H_5)_2SnCl_2 \cdot PBI$	<1	<1	>1	<1	<2
$(C_2H_5)_2SnCl_2 \cdot phen$	>1	<1	<1	<1	<2
$(C_2H_5)_2SnBr_2 \cdot PBI$	<1	<1	<1	<1	<1.9
$(C_2H_5)_2SnBr_2 \cdot phen$	>2.5	<1	<1	<1	<1.9
$(C_6H_5)_2SnCl_2 \cdot PBI$	<1	<1	<1	<1	<0.8
$(C_6H_5)_2SnCl_2 \cdot phen$	1	<1	4	<1	<2.3
$(C_6H_5)_2SnBr_2 \cdot PBI$	1	<1	<1	<1	<0.9
$(C_6H_5)_2SnBr_2 \cdot phen$	<1	<1	<1	<1	<2.5
$(C_6H_5)_2SnCl_2 \cdot 2DMSO$	<1	<1	<1	<1	<2.5

[a] S.I. = ratio of CD_{50} to ED_{50}.

pharmaceutical will develop. There is still considerable interest in the exploration of the antitumour properties of tin compounds and a highly effective tin derivative may yet be found. Finally since tin compounds and in particular the organotins exhibit such a wide variety of biological activity, they ought to be considered for inclusion in other pharmaceutical screening programmes.

ACKNOWLEDGEMENTS

Prof. Marcel Gielen is thanked for his invitation to present this paper, as are the Conference sponsors; Nato Scientific Affairs Division, Brussels, Belgium, for financing my attendance. The International Tin Research Institute, Uxbridge is thanked for permission to publish this paper. Thanks also go to the following ITRI staff; Dr Peter Smith for his comments on the manuscript; Mrs Karen Skiggs for her typing services; and Mr Greg Collins, Mrs Kathy Anson and Miss Janet Ratcliffe for assistance with photographs and slides.

REFERENCES

Arakawa Y and Wada O (1984) Inhibition of carrageenan edema formation by organotin compounds. Biochem Biophys Res Comm 125: 59-63

Arakawa Y (1986) Side effect of organotin compound possessing antitumour activity. 3rd Internat Symp Tin Malign Cell Growth, Padua, Sept 5-6: 8

Atassi G (1985) Antitumour and toxic effects of silicon, germanium, tin and lead compounds. Rev Si Ge Sn Pb Comp 8: 219-235

Barbieri R, Pellerito L, Ruisi G and Lo Giudice MT (1982) The antitumour activity of diorganotin(IV) complexes with adenine and glycylglycine. Inorg Chim Acta 66: L39-L40

Ben-Hur E and Rosenthal I (1985a) The phthalocyanines: a new class of mammalian cells photosensitizers with a potential for cancer phototherapy. Int J Radiat Biol 47: 145-147

Ben-Hur E and Rosenthal I (1985b) Factors affecting the photo-killing of cultured Chinese hamster cells by phthalocyanines. Radiat. Res. 103: 403-409

Berglund L, Angelin B, Blomstrand R, Drummond G and Kappas A (1988) Sn-Protoporphyrin lowers serum bilirubin levels,

decreases biliary bilirubin output, enhances biliary heme excretion and potentially inhibits hepatic heme oxygenase activity in normal human subjects. Hepatol 8: 625-631
Blunden SJ, Cusack PA, Smith PJ and Barnard PWC (1983) Studies on the interaction of inorganic tin compounds with methyl 4,6-O-benzylidene-α-D-glucopyranoside and related molecules. Inorg Chim Acta 72: 217-222
Blunden SJ, Cusack PA and Hill R (1985) The Industrial uses of tin chemicals. Royal Society of Chemistry, London
Brown NM (1972) The effect of two organotin compounds on C3H-strain mice. PhD Thesis, Clemson University
Bulten EJ and Budding HA (1981) Tin compounds for the treatment of cancer. Brit Pat 2,077,266
Cardarelli NF, Cardarelli BM and Marioneaux M (1983) Tin as a vital trace nutrient. J Nut Growth Cancer 1: 181-194
Cardarelli NF, Quitter BM, Allen A, Dobbins E, Libby EP, Hager P and Sherman LR (1984a) Organotin implications in anti-carcinogenesis.1. Background and thymus involvement. Austral J Exp Biol Med 62: 199-208
Cardarelli NF, Cardarelli BM, Libby EP, and Dobbins E (1984b) Organotin implications in anticarcinogenesis.2. Effects of several organotins on tumour growth rate in mice. Austral J Exp Biol Med 62: 209-214
Cardarelli NF (1985) Tin steroids as anticancer agents. 2nd Internat Symp Tin Malignant Cell Growth, Scranton, May 20-22
Cardarelli NF and Kanakkanatt SV (1985) Tin steroids and their use as antineoplastic agents. US Patent 4,541,956
Cardarelli NF and Kanakkanatt SV (1987) Tin steroids and their uses. US Patent 4,634,693
Carrara M, Zampiron S, Voltarel G, Sindellari L and Trincia L (1988) Inhibitory properties of tin(IV) diethyl-dithiocarbonates on tumoral cells growth. Pharmacol Res Commun 20: 611-612
Chan W-S, Marshall JF, Svenson R, Phillips D and Hart IR (1987) Photosensitising activity of phthalocyanine dyes screened against tissue culture cells. Photochem Photobiol 45: 757-761
Cleare MJ (1974) Transition metal complexes in cancer chemotherapy. Coord Chem Rev 12: 349-405
Cornelius CE and Rodgers PA (1984) Prevention of neonatal hyperbilirubenemia in Rhesus monkeys by tin protoporphyrin. Pediatr Res 8: 728-730
Cornelius CE, Rodgers PA Bruss ML and Ahlfors CE (1985) Characterisation of a Gilbert-like syndrome in squirrel monkeys (Saimiri sciureus). J Med Primatol 14: 59-74
Crowe AJ (1980) Synthesis and studies of some biologically active organotin compounds. PhD Thesis, London University
Crowe AJ and Smith PJ (1980) Dialkyltin dihalide complexes: A new class of metallic derivatives exhibiting antitumour activity. Chem Ind 200-201
Crowe AJ, Smith PJ and Atassi G (1980) Investigations into the antitumour activity of organotin compounds.1. Diorganotin dihalide and dipseudohalide complexes. Chem Biol Interact 32: 171-178
Crowe AJ, Smith PJ and Atassi G (1984a) Investigations into

the antitumour activity of organotin compounds.2. Diorganotin dihalide and dipseudohalide complexes. Inorg Chim Acta 93: 179-184

Crowe AJ, Smith PJ, Cardin CJ, Parge HE and Smith FE (1984b) Possible pre-dissociation of diorganotin dihalide complexes. Relationship between antitumour activity and structure. Cancer Lett 24: 45-48

Crowe AJ (1987) The chemotherapeutic properties of tin compounds. Drugs of the Future.3: 255-275

Crowe AJ (1988a) The antitumour activity of tin compounds. In: Gielen M (ed) Metal based Antitumour Drugs. Freund, Tel Aviv, p 103

Crowe AJ (1988b) The pharmaceutical applications of tin compounds. Proceeding 31st German Tin Day, Düsseldorf, 26th April, 56-70

De Clerq L, Willem R, Gielen M and Atassi G (1984) Synthesis, characterisation and antitumour activity of bis(polyfluoroalkyl)tin dihalides Bull Soc Chim Belg 93: 1089-1097

Delaney JK, Mauzerall D, Drummond G and Kappas A (1988) Photophysical properties of tin-protoporphyrins: potential clinical applications. Pediatrics 81: 498-504

Deutsch E, Elder RC, Lange BA, Vaal MJ and Lay DG (1976) Structural characterisation of a bridged ^{99}Tc-Sn-dimethylglyoxime complex: Implications for the chemistry of ^{99m}Tc-radiopharmaceuticals prepared by the Sn(II) reduction of pertechnetate. Proc Natl Acad Sci USA 73: 4287-4289

Drummond GS and Kappas A (1981) Prevention of neonatal hyperbilirubenaemia by tin protoporphyrin IX, a potent competitive inhibitor of heme oxidation. Proc Natl Acad Sci USA 78: 6466-6470

Drummond GS and Kappas A (1982) Chemoprevention of neonatal jaundice: Potency of tin-protoporphyrin in an animal model. Science 217: 1250-1252

Eng G and Engle TW (1987) Synthesis and potential anticancer activity for diaminoalkyl complexes of tin halides. Bull Soc Chim Belg 96: 69-70

Evans CJ and Karpel S (1985) Organotin compounds in modern technology. J. Organometallic Chemistry Library. Elsevier, Amsterdam

Gielen M, Jurkschat K and Atassi G (1984) Bis(halophenylstannyl)methanes: New organotin compounds exhibiting antitumour activity. Bull Soc Chim Belg 93: 153-155

Gielen M, Joosen E, Mancilla T, Jurkschat K, Willem R, Roobol C, Bernheim J, Atassi G, Huber F, Preut H and Mahieu B (1987) Diorganostannylene derivatives of 2,6-pyridine dicarboxylic acid: Synthesis, spectroscopic characterization, X-ray structure amalysis, in vitro and in vivo antitumour activity Main Group Metal Chem 10: 147-167

Gielen M, Willem R, Mancilla T, Ramharter J and Joosen E (1988) In: Zuckerman JJ (Ed) Tin and malignant cell growth. CRC Press, Cleveland

Haiduc I, Silvestru C and Gielen M (1983) Organotin compounds: New organometallic derivatives exhibiting antitumour activity. Bull Soc Chim Belg 92: 187-189

Hefferren JJ (1963) Qualitative and Quantitative tests for stannous fluoride. J Pharm Sci 52: 1090-1096

Howell CL, Gish CW, Smiley RD and Muhler JC (1955) Effect of topically applied stannous fluoride on dental caries experience in children. J Am Dent Assoc 50: 14-17

Huber F, Roge G, Carl L, Atassi G, Spreafico F, Filippeschi S, Barbieri R, Silvestri A, Rivarola E, Ruisi G, Di Bianca F and Alonzo G (1985) Studies on the antitumour activity of di- and tri-organotin (IV) complexes of amino acids and related compounds, of 2-mercptoethane sulphonate, and of purine-6-thiol. J Chem Soc Dalton Trans 523-527

Iwai H, Manabe M, Ono T and Wada O (1979) Distribution, biotransformation and biological half-life of tri-, di-, and mono-butyltin in rats. J Toxicol Sci 4: 285

Jang JJ, Takahashi M, Furukawa F, Toyoda K, Hasegawa R, Sato H and Hayashi Y (1986) Inhibitory effect of dibutyltin dichloride on pancreatic adenocarcinoma development by N-nitrosobis (2-oxopropyl)amine in the Syrian Hamster. Jap J Cancer Res (Gann) 77: 1091-1094

Jiazhu W, Jingshuo H, Liyiao H, Dashuang S and Shengzhi H (1988) Antitumour activity of organotin compounds. Reaction, synthesis and structure of Et_2SnCl_2(phen) with 5-fluorouracil. Inorg Chim Acta 152: 67-69

Kanisawa M and Schroeder HA (1967) Effect of arsenic, germanium, tin and vanadium on spontaneous tumours in mice. Life term studies. Cancer Res 27: 1192-1195

Kanisawa M and Schroeder HA (1969) Life term studies on the effect of trace elements on spontaneous tumours in mice and rats. Cancer Res 29: 892-895

Kappas A, Drummond GS, Simionatto CS and Anderson KE (1984) Control of heme oxygenase and plasma levels of bilirubin by a synthetic heme analogue, tin-protoporphyrin. Hepatol 4: 336-341

Kappas A, Drummond GS, Manola T, Petmezaki S and Valaes T (1988) Sn-Prototporphyrin use in the management of hyperbilirubinemia in term newborns with direct Coombs-positive ABO incompatibility. Pediatrics 81: 485-497

Keller HJ, Keppler B, Kruger U and Linder R (1982) Antineoplastic effect of metal complexes and their use in medicine. Eur Pat 49,486

Keyes JM Jr, Carey J, Moses D and Beierwaltes W (1973) CRC Manual of Nuclear Medicine Procedures, 2nd Ed. CRC Press, Cleveland

Konig KG (1959) Dental caries and plaque accumulation in rats treated with stannous fluoride and penicillin. Helv Odontol Acta 3: 39-41

Köpf H and Köpf-Maier P (1983) Tumour inhibition by metallocene dihalides of early transition metals. In: Lippard SJ (Ed) Platinum, gold and other metal chemotherapeutic agents: Chemistry and biochemistry. ACS Symp Ser 209

Köpf-Maier P, Janiak C and Schumann H (1988) Monomeric airstable metallocenes of main-group elements as antitumour agents. Inorg Chim Acta 152: 75-76

Krause E (1929) An effective treatment of experimental mousecancer using organolead compounds. Ber 62: 135-137

Lim JKJ (1970) Precipitate-free, dilute aqueous solutions of stannous fluoride for topical application.1. Simple and mixed mediums. J Dent Res 49: 760-767

Meinema HA, Liebregts AMJ, Budding HA and Bulten EJ (1985) Synthesis and evaluation of organometal-based antitumour agents of germanium and tin. Rev Si Ge Sn Pb Comp 8: 157-168

Muhler JC and van Huysen G (1947) Solubility of enamel protected by sodium fluoride and other compounds. J Dent Res 26: 119-127

Moreno G, Pottier RH, Truscott TG (1988) Photosensitisation. Springer-Verlag, Berlin Hiedelberg New York

Morgan AR, Garbo GM, Keck RW and Selman SH (1987a) Tin(IV) etiopurpurin dichloride: an alternative to DHE? Proc SPIE-Int Soc Opt Eng 847: 172-9

Morgan AR, Garbo GM, Keck RW and Selman SH (1987b) Hydrophobic versus hydrophilic drugs for PDT. Proc SPIE-Int Soc Opt Eng 847: 180-186

Morgan AR, Kreimer-Birnbaum M, Garbo GM, Keck RW and Selman SH (1987c) Purpurins: Improved photosensitizers for photodynamic therapy. Proc SPIE-Int Soc Opt Eng 847: 29-35

Morgan AR, Garbo GM, Keck RW and Selman SH (1988) In vivo cytotoxicity of metallopurpurins to bladder tumours. In: Moreno G, Pottier RH and Truscott TG (eds) Phototsensitisation, Springer-Verlag, Berlin Hiedelberg New York, p 495

Nagae H, Keino H, Watanabe K, and Kashiwamata S (1988) Pharmacological and biological effects of tin-protoporphyrin on neonatal hyperbilirubinemic Gunn rats. Pediatr Res 24: 209-12

Peters W, Trotter ER and Robinson BL (1980) The experimental chemotherapy of leishmaniasis, VII. Drug responses of L. major and L. mexicana amazonensis, with an analysis of promising chemical leads to new antileishmanial agents. Ann Trop Med Parasitol 74: 321-335

Pottier R (1988) Past, present and future of photosensitizers. In: Moreno G, Pottier RH and Truscott TG (eds) Photosensitisation. Springer-Verlag, Berlin Hiedelberg New York, p 1

Prestayko AW, Crooke ST and Carter SK (Eds) (1980) Cisplatin: Current status and New developments. Academic, New York

Reddi E, Lo Castro G, Romandini P and Jori G (1986) Preliminary studies on the use of Zn-phthalocyanines in photodynamics therapy. Abstract from: Porphyrin Photosensitization Workshop, Los Angeles 26-27 June

Rideout D, Kappas A and Drummond GS (1987) Tin diiododeuteroporphyrin and therapeutic use thereof. US Pat 4,668,670

Robins RK (1986) Synthetic antiviral agents. Chem Eng News, Jan 27: 28-40

Ruisi G, Silvestri A, Lo Giudice MT, Barbieri R, Atassi G, Huber F, Grätz K and Lamartina L (1985) The antitumour activity of di-n-butyltin(IV)glycylglycinate, and the correlation with the structure of dialkyltin(IV)glycylglycinates in solution studied by conductivity measurements and by infrared, nuclear magnetic resonance and Mössbauer spectroscopic methods, J Inorg Biochem 25: 229-245

Sacerdoti D, Escalante B, Abraham NG, McGiff JC, Levere RD and Schwartzman ML (1989) Treatment with tin prevents the development of hypertension in spontaneously hypertensive rats. Science 243: 388-390

Sadler PJ (1982) Inorganic pharmacology. Chem Brit 18: 182-

184, 188
Sassa S, Drummond GS, Bernstein SE and Kappas A (1983) Tin protoporphyrin suppresion of hyperbilirubenemia in mutant mice with severe hemolytic anemia. Blood 61: 1011-1013
Sassa S, Drummond GS, Bernstein SE and Kappas A (1985) Long term administration of massive doses of Sn-protoporphyrin in anemic mutant mice. J Exp Med 162: 864-876
Saxena AK, Koacher JK, Tandon JP and Das SR (1982) Studies of organotin-Schiff base complexes as new potential amebicidal agents. J Toxicol Environ Health 10: 709-715
Saxena A and Tandon JP (1983) Antitumour activity of some diorganotin and tin(IV) complexes of Schiff bases. Cancer Lett 19: 73-76
Scherer Laboratories Inc (1981) Stannous fluoride as a plaque inhibitor: Human clinical studies. 'Gel-Kam' Prevent Dent Rev 4(2)
Shannon IL (1970) Antisolubility effects of acidulated phosphofluoride and stannous fluoride in the treatment of crown and root surfaces. Austral Dent J 16: 240-242
Shannon IL and Wightman JR (1970) Treatment of root surfaces with a combination of acidulated phosphatofluoride and stannous fluoride. J Louisiana Dent Assoc 28: 14-17
Sherman LR, Coyer MJ and Huber F (1987) Synthesis of tributyltin taurocholate, taurodeoxycholate and glycocholate. Appl Organomet Chem 1: 355-358
Sherman LR and Huber F (1988) Relationship of cytotoxic groups in organotin molecules and the effectiveness of the compounds against leukaemia. Appl Organomet Chem 2: 65-72
Sherman SE, Gibson D, Wang AH-J and Lippard SJ (1985) X-ray structure of the major adduct of the anticancer drug cisplatin with DNA: $\underline{cis}$-[Pt(NH$_3$)$_2${d(pGpG)}] Science 230: 412-417
Sisson TRC, Drummond GS, Samonte D, Calabio R and Kappas A (1988) Tin-protoporphyrin blocks the increase in serum bilirubin levels that develops postnatally in homozygous Gunn rats. J Exp Med 167: 1247-52
Stout DL and Becker FF (1988) The effects of tin-protoporphyrin administration on hepatic xenobiotic metabolising enzymes in the juvenile rat. Drug Metab Dispos 16: 23-26
Svatun B and Attramadal A (1978) The effect of stannous fluoride on human plaque acidogenicity in situ (Stephan curve). Acta Odontol Scand 36: 211-218
Sykes AG (1988) Reactions of complexes of platinum metals with bio-molecules. Plat Met Rev 32: 170-178
Takahashi M, Furukawa F, Kokubo T, Kurata Y and Hayashi Y (1983) Effect of dibutyltin dichloride on incidence: of pancreatic adenocarcinoma induced in hamsters by a single dose of N-nitrosobis(2-oxopropyl) amine. Cancer Lett 20: 271-276
Wada O and Manabe S (1986) Biochemistry, pharmacodynamics and kinetics of tributyltin compounds. Proc ORTEPA-Workshop, Berlin, May 15-16: 113-121
Ward SG, Taylor RC and Crowe AJ (1988) The *in vitro* antiherpes activity of some selected antitumour organotin compounds. Appl Organomet Chem, 2: 47-52
Ward SG, Taylor RC, Crowe AJ, Balzarini J and De Clerq E (1989) The broad spectrum *in vitro* antiviral activity of some

selected antitumour-active organotin complexes. Appl Organomet chem, submitted for publication

Williams HJ, Shannon IL and Stevens FD (1974) The treatment of intact root surfaces with combinations of fluoride. J Am Soc Prevent Dent 4: 40-44

Wissel PS, Galbraith RA, Sassa S and Kappas A (1988) Tin-protoporphyrin inhibits heme oxygenase and prevents the decline in hepatic heme and cytochrome P-450 contents produced in nude mice by tumour transplantation. Biochem Biophys Res Commun 150: 822-827

Wright PS (1980) The effect of soft lining materials on the growth of Candida albicans. J Dentistry 8: 144-151

van Lier JE, Brasseur N, Paquette B, Wagner JR, Ali H, Langlois R and Rousseau (1988) Phthalocyanines as sensitisers for photodynamic therapy of cancer. In: Moreno G, Pottier RH and Truscott TG (eds) Photosensitisation. Springer-Verlag, Berlin Hiedelberg New York, p 435

THE ROLE OF TIN HORMONES IN SENESCENCE: A HYPOTHESIS

Nate F. Cardarelli
Engineering and Science Technology Division
The University of Akron
Akron, Ohio 44325-6104, U.S.A.

This symposium, the fifth of a series, has been initiated and maintained by our common interest in the effects of tin compounds on elements of the immune system and their potential usage of such materials in cancer therapy. This work has been well documented in the earlier Proceedings and journal articles, and will generally only be mentioned here by citation. It is the intent of this report to assess the "fit" of the subject data into a conceptual framework termed the "Life Pattern Hypothesis of Senescence". While past attention has been riveted on tin agents as potential therapeutic or prophylactic agents for malignant disease, the ancillary role of endogenous tin in mammalian development and aging has been neglected. This document serves as an attempt to integrate the dual roles of tin within the conceptual schema of the hypothesis.

BACKGROUND: IMMUNE INVOLVEMENT

In the early years of this decade, it was discovered that exogenous tin accumulates in the thymus gland (Cardarelli et al 1984a; Cardarelli 1986a; Cardarelli

1989a)i Since the thymus had never been recognized as a xenobiotic depot, the presence of tin in relatively high concentration and its propensity to home in on this site suggested physiological utility. Measuring tin content in the thymi of humans and other species demonstrated an intraspecies variation as well as similar age-dependent concentration levels within the same species, barring pathological intervention or genetic alterations such as seen in laboratory animals (Bilgicer et al 1985; Cardarelli 1986a; Cardarelli 1989a; Cardarelli et al 1984b). In elaborate time profile studies using a tin isotope and measuring tissue and excretia concentrations by gamma dosimetry results suggested that (Cardarelli et al 1985; Cardarelli 1986; Peterson et al 1986):

1) Xenobiotic tin leaves the intestinal lumen through the villi entering the mesenteric lymph nodes - and from there to the mouse thymus via lymphatic circulation.

2) Thymus dwell time is six to eight hours, then label moves outward first to peripheral lymphatic tissues and then to all the organs.

3) Blood tin content is extremely low in outbred cancer refractory mice, but much higher in tumor bearing or tumor prone inbred strains.

4) The circulatory path for tin administered to cancer prone mice is significantly different, with reduced or absent deposition in the thymus.

Such data suggested a biosynthesis capability in the refractory mouse and its absence in the high cancer incident strains.

The propensity of the thymus to accumulate exogenous tin has been amply demonstrated by a number of studies indicating that specific alkyltins cause reversible dose-dependent lesions in this gland (Miller et al 1984; Penninks et al 1986; Seinen et al 1976). Inorganic tin and natural tin compounds of plant origin are not so implicated (Cardarelli 1988a). As Paracelsus, the "Father of Pharmacy", remarked long ago - "All things are poisons, the difference is in the dosage" (author's translation). Studies with a number of lower di- and trialkyltins indicate the presence of a threshold toxicity. An acute dosage above threshold causes tissue destruction within a given organ (thymus, hippocampus, etc.) while multiple subthreshold dosages, even though the aggregate be above this limit, are morphologically innocuous. There is thus the implication that either the agent is metabolized at a set rate or the effect is indirect - perhaps mediated by the adrenal cortex (Cockerill et al 1987). The known ill effects of certain alkyltins on the thymus buttresses data indicating that tin accumulates therein. It has been shown that in the thymus and bursa chemically induced involution is dose-dependent and does not occur under 2 ppm dibutyltin dosages (Gota-Socaciu et al 1986). The same study also demonstrated that this agent acts directly on the thymus and not through the adrenals.

Various alkyltin halides affect several components of the immune system through direct detrimental activity on the lymphocyte and indirectly by degradation of the thymus resulting in loss of cortical T cells and decreased hormonal secretions (Cardarelli 1989b). However, elemental tin, inorganic tin, and numerous organotin compounds act as immune stimulants - although specific mechanisms remain to be elucidated (Cardarelli 1989b; Levine 1986; Levine et al 1982). Tin powder causes a dramatic lymphoadenopathy in draining lymph nodes and marked increases in immunoglobulin secreting plasma cells (Levine et al 1987). Tin is not an antigen - therefore why an antibody response?

Numerous reports, reviewed elsewhere, indicate that tin retards tumor growth rate and reduces the incidence of spontaneous neoplasia in test animals (Cardarelli 1989a; Penninks et al 1986). Organotin compounds have shown both therapeutic and prophylactic merit in this regard (Cardarelli 1986c; Cardarelli et al 1984c; Carrara et al 1988; Crowe et al 1980; Gielen 1986). Labelled tin ion or organotin compounds presented to mice appear to preferentially move to the tumor, extensively decorating the membrane, though with relatively slight penetration (Cardarelli 1986a; Cardarelli 1986c).

Analysis of cancerous human breast and lung tissue showed low tin content in the tumor but relatively high concentrations in surrounding nonmalignant tissues (Bilgicer et al 1985, Sherman 1986). It has been reported that calcifications associated with human mammary cancer contain a significantly high tin concentration, second only to calcium content as analyzed with a scanning electron microscope equipped with a backscatter electron detector and an X-ray dispersive spectrometer (Galkin et al 1987). The evidence, admittedly sparse, for tin having a prophylactic effect on tumor incidence has been discussed elsewhere (Cardarelli et al 1984b; Cardarelli et al 1984c; Cardarelli 1986c).

From such data, it was hypothesized by the author that the thymus processes tin into one or more factors that either stimulate specific anticancer elements of the immune system, or act as a chalone within the genome of malignant cells. The presence of a potent antiproliferative steroid of unknown structure in the thymus, led to the author's belief that this "S" factor was a tin bearing cholesterol derivative (Cardarelli et al 1988b). Consequently a series of tin steroids were synthesized and evaluated against mouse adenocarcinoma _in vivo_, and human epidermoid tumor _in vitro_ (Cardarelli et al 1985a; Cardarelli 1988b; Cardarelli 1989b). Four compounds demonstrated marked tumor reduction properties _in vivo_, and in some instances transplants were totally necrotic after a set o.p. administrative period. _In vitro_ studies with human and mouse tumor cells indicated

ID_{50} values similar to those observed with in-use anticancer therapeutic agents (Cardarelli 1988b). Results do not confirm that the "S" factor is a tin steroid, only that tin steroids are anticarcinogenic. However, in a recent study of tin content and antiproliferative potential of a series of thymic extracts following the Potop et al. isolation scheme, a close correspondence existed between the tin content of each fraction and anticancer effectiveness (Cardarelli et al 1985a; Cardarelli et al 1988; Potop et al 1970).

BACKGROUND: GEROPROTECTIVE EFFECTS OF TIN

Schroeder and his colleagues demonstrated some years ago that rodents presented with a high tin diet not only showed a marked decrease in malignancies and other lesions, but also a significant extension of mean lifetime; up to 22% in some instances (Cardarelli 1986d; Schroeder et al 1967). It has also been observed that tin needles inserted intrathoracically in marsh mice prolonged average life span (Bischoff et al 1976). However, studies on human tin consumption related to disease incidence and lifespan are conspicuously absent. The mortality rate of tin miners reportedly is reduced 24% at any given age (Robertson 1964). Unfortunately this report lacks sufficient data to support the claim.

CAPITULATION

It can be reasonably postulated on the basis of the background information provided herein and the much larger data pool in the cited references, that exogenous tin is 1) bioprocessed into one or more factors or hormones that function within the immune compartment and 2) tin prolongs life in some unknown manner, directly or indirectly through its immune activity. While these claims have been advanced in previous symposia and published in a number of journal articles, the conceptual framework has not been elucidated. That is: do the above conclusions and their supporting data conform to the gerontological paradigm? In order to meet that question, this report examines the current theory of aging and its extension through what is now termed the "Life Pattern Hypothesis of Senescence".

ONTOGENY OF THE IMMUNE SYSTEM

The vertibrate immune system has two prime functions; detection and destruction of exogenous invaders such as bacteria, viruses, antigens, etc.; and maintenance of homeostasis. The second involves regulation of cell differentiation, maturation, mitotic events, and atrophication; and the control of numerous organismic biochemical processes and pathways. Unless properly controlled, a cell has but two possible destinies - death, or unlimited growth through mitosis. This latter possibility we term malignancy; and though the cell achieves immortality in the sense of infinite division potential, it ultimately destroys the host organism.

Individual immune capacity developes in well recognized phases with the thymus gland as the primordium. The thymic anlagen is morphologically distinct in the human by the tenth gestational week. In utero division into two distinct compartments - medulla and cortex are observed. The cortical component gradually fills with prelymphocytes and through a series of antigen/hormone interactions, mature lymphocytes develop. The medulla is comprised of a number of epithelial type cells some or all of which are secretory. A number of thymic hormones and their involvement in the immune system maturation have been amply described in an extensive literature reviewed elsewhere (Cardarelli 1986e; Cardarelli 1989a, 1989b; Kendall 1981; Luckey 1973; Zeppezauer et al 1988). The maturation process radiates outward from thymus to spleen, lymph nodes, and other peripheral lymphatic tissues. Mediation of this process by the thymus is critical during in utero and neonatal human life.

Man, like many rodent species, is a secondarily altricial animal; born as an embryo with numerous body processes in an immature state (Gould 1975). To be born with the same degree of maturation seen in precocial animals, man's gestation period would be nearly 18 months! The human immune system at parturition lacks vital competance, the infant relies on maternal antibodies received in utero and bolstered by further antibody input via maternal colostrum and lactates. Thymus essentiality exists well into the fourth postnatal month and possibly longer. The thymus achieves maximum size by adrenarche and is in noticeable decline several years later at gonadarche. Involution lasts throughout life marked by a steady depletion of cortical cells, atrophication in the medulla, and decline in hormonogenesis. Transitory reversible involution is noted with sickness, pregnancy and malnutrition (Baarlen et al 1988; Cardarelli 1989b; Luckey 1973) Athymic neonates, man or mouse, cannot long survive in the post partem environment. Also, the thymi of growth-retarded infants dying at or about parturition were found considerable below normal relative to body weight (Hartge et al 1987).

IMMUNE THEORY OF AGING

The current, in vogue aging theory essentially relates loss of immune capability to senescence (Walford 1969; Walford 1983). Involution of the thymus, followed by a more gradual atrophication of other lymphoid tissue results in the decline of immune function and associated losses in homeostatic mechanisms. The varied organismic alterations accompanying senescence can be reasonably associated with loss in immune potential. Superficial changes in the skin, vision losses, muscle wasting, shifts in mental acuity, etc. and the pathologies seen with advancing age can be linked to immune system failure. Death, barring accidental trauma, is associated with four primary pathological conditions: cardiovascular failure, autoimmune disease, infection, and malignancy. Each category has been directly related to decreased immune function.

The Immune Theory sets forth the mechanism by which immune loss translates into aging phenomena. It cannot explain processes at either end of the cause and effect chain. At the cellular level, specific gene expression is of necessity changed in order that the underlying biochemical processes supporting lymphatic cells are altered in such a way as to insure immune decline. Gene expression may be turned off, or conceivably turned on. The probabilities lie in terminating critical gene expression through alteration of the histone-DNA linkage sites. (It is notable that one or more thymus hormones are macromolecules containing histone segments (Zeppezauer et al 1988). However, the "Immune Theory" as formulated, cannot encompass or somehow explain the mechanism involved at the genome level. More importantly, the immune theory is mute regarding the verae causae of the processes leading to loss of immune capability. This and other shortcomings of this major, now well accepted, aging theory have been discussed in considerable detail elsewhere (Cardarelli 1989b, 1990; Goidl 1987; Walford 1969, 1983).

THYMUS-PINEAL AXIS

The thymus does not program itself, but rather relies on

humoral mediation and possibly direct neural communication with higher centers. Although thymus-pituitary, thymus-adrenal, and thymus-thyroid axes exist, analysis of the extensive available data led the author to also conclude that an active thymus-pineal linkage is present in the fetal period and, in altricial animals, also during early neonatal life (Cardarelli 1989a). In the Life Pattern hypothesis, one major tenet is that the pineal gland programs the thymus. This concept is based essentially on the priority of the pineal in time; it is present by the 33rd gestational day in man, and displays an early transitory immune function. During the interval between the second month *in utero* to about twelve months post partem, the pineal undergoes marked changes wherein immune function is gradually lost. The early pineal possesses lymphocytes, secretes humoral substances that affect immune development, and contains growth regulatory factors (Anisimov et al 1982; Blask 1984; Damian 1985; Lapin 1976; Tapp 1980; Uede et al 1981). Antiblastic compounds have been isolated from the pineal and its early extirpation leads to a marked enhancement of cancer incidence and growth. The *in utero* pineal displays a transitory nerve linkage with higher centers from about the third to sixth gestational month. The "Nervus Pinealis" is found in the human fetus of 56mm to 169mm crown to rump length, and in sheep and rabbit fetuses (Møller 1979). Also, profound histological and morphological changes are noted in the pinealocytes and accessory cells during the critical *in utero* - neonatal period (Kurumodo et al 1976; Vollrath 1986).

It is well recognized that the pineal is involved in the biological clocking mechanisms that govern a number of circadian, and circannual rhythms (Pengelley 1974; Reiter, 1981). In numerous animal species, the pineal mediates seasonal breeding through release of humoral factors at the appropriate time. The timing is based upon the exogenous photoperiod - i.e. day length, an astrophysical parameter (Gwinner 1981). The timing of human puberty appears to be a pineal function. In specific cancerous conditions where the pineal's hormonal function is compromised, precocious puberty occurs, whereby, the male or female shows sexual maturation and the ability to reproduce at extremely young ages (Waldhauser et al 1986). For instance, menstruation at age three, and the earliest confirmable birthing at age five years, six months (Escomel 1939)! In general, the thymus and pineal act in concert to regulate puberty, with the former advancing and the latter retarding timing of the event (Besedovsky et al 1974).

Both thymus and pineal are involved in somatic and germ cell growth and maturation during interuterine life and for a species-dependent period after birth (Cardarelli 1986d; Cardarelli 1989b; Kendall 1981). Pinealectomy of the adult rat leads to deficits in immune and other functions characterized by thymus atrophy, malignancy, and metabolic alterations (Barath et al 1974). Human, bovine and rat thymocytes secrete TGP (thymocyte growth peptide) which recruits immature thymocytes into S phase of the cell cycle (Ernstrom et al 1988).

LIFE PATTERN HYPOTHESIS

Basically the life pattern hypothesis conceives a pineal to thymus flow of information in early life that programs the thymus for its role in immune system maturation. The pineal functions as a neuroendrocrine transducer throughout life providing hormonal pulses at intervals dictated by nerve signals from known areas of the hypothalamus (Goldman et al 1983; Maestroni et al 1988). The major pacesetters, or oscillators, are the suprachiasmatic nucleus (Inouye et al 1979), paraventricular nucleus (Baertschi et al 1982), supraoptic nucleus (Leng et al 1982), and ventromedial nucleus (Brainard et al 1986). Each receives cyclical exogenous signals, or zeitgebers, that trigger a pulsatile neuroresponse governing a host of periodic human functions. There are a number of rhythms established such as ultradian (less than 24 hours) - e.g. the 90 minute pituitary growth hormone release pulse; and circadian (around 24 hours) e.g. sleeping, eating, excretion, blood pressure, adrenocortical steroid release, etc. Infradian cycles are those whose period is greater than one day; such as the septradian testosterone rhythm or the ovulation cycle. Circannual rhythms follow the yearly course of the sun, and in humans, well defined annual immune and cardiac rhythms are observed. It may be that when distinct circannual and circadian rhythms oscillate at acrophase, life events, such as gonadarche or the climacteric occur. Biological rhythms are fully established (or "entrained") by the first or second year of infancy. Rhythmic disturbances lead to affective disorders ranging from jet lag to schizophrenia, including seasonal affective disorders, manic-depressive psychoses, Ondines curse, the fatal "night terrors", possibly Alzheimer's disease, and perhaps the sudden infant death syndrome (Cardarelli 1989a, 1990). Zeitgebers include not only photoperiod which may be of lesser importance, but also ionizing radiation of solar origin, geomagnetism, gravitational fields (including sun, earth, and moon) and terrestrial electric fields. Hypothalamic oscillators vibrate in accordance with the cyclicity of environmental parameters (Brainard et al 1986; Cremer-Bartles 1984; Michaelson 1986; Osipov et al 1971).

Interference with the zeitgebers alters the life pattern of the individual. Completely blind humans cannot detect daily photoperiod through the optical tract. Such individuals free run, when deprived of sensory cues as to day/night conditions, at 24.9 hour cycles (Miles et al 1977). This change in subjective day length translates into a mean lifespan some 3.5 years greater than normal, after adjusting for the major confounder - accidental death (Zacharias et al 1964). Females blind from birth reach puberty about one year early and menopause around two years later than average (Magee et al 1970).

A partial description of the Life Pattern modification of the Immune Theory of Aging has been partially described elsewhere, (Cardarelli 1989a) and a complete treatise is in preparation (Cardarelli 1990). At this point of the narrative, I wish to consider the potential role of circulating tin bearing factors in the concepts summarized above.

TIN DEPRIVATION AND LIFE SPAN

The phenomenon that first drew my attention to tin and the aging process was the similarity of symptoms between thymectomized rodents and those deprived of dietary tin. Neonatally thymectomized altricial mammals and young rats isolated from contact with tin exhibit the following syndrome (Cardarelli 1986f; Miller 1964; Schwarz 1974; Schwarz et al 1970):

>Failure to grow (Dwarfism)
>Muscle wasting (Cachexia)
>Alopecia
>Weight Loss
>Lack of energy
>Lack of tonicity
>Seborrhea

Life span is dramatically reduced to about three months in thymectomized mice, rats, dogs, and other Altricial species. While Schwarz and his colleagues never carried tin deprivation to the terminal stage, based on the observed rate of weight loss, one can estimate a three- to four-month lifespan. It has also been reported that rodents on a dramatically reduced level of dietary tin exhibit thymic involution (Sherman et al 1986b).

Short lived rodents, such as the Snell-Bagg, Ames, grey-lethal and "nude" mouse; and those that spontaneously develop cancer, e.g. the AKR leukemic strain, show gross thymus aplasia - and in the latter, very low thymic tin content. The above syndrome applies equally well to a number of mouse, rat, and rabbit dwarf strains - where gross thymus abnormalities are coupled with about a 10 to 15% normal life span. In some instances, the thymus is nearly or wholly absent, such as in the several "nude" rodent strains, and the syndrome terminating in death occurs as pineal immune capacitance declines after birth. In general, and unlike thymectomized mice, the transplantation of thymus tissue does not reverse the syndrome - because, according to the hypothesis, the pineal-thymus linkage developed in utero is severed usually by two weeks post natal and cannot be re-established through this expedient.

The Snell-Bagg dwarf mouse with an average 45 to 65 day lifespan, is of interest. At birth the thymus is morphologically and histologically normal, body weight and growth curves fall within the normal range. At about the tenth post natal day a rapid involution of the thymus commences, weight gain is arrested, wasting initiates, secondary myxedemia is pronounced, and BMR is only 40% of normal (Bartke 1964; Duquesnoy et al 1970). Interestingly, dwarf mutant life span is extended by administering mouse milk after weaning.

The accelerated aging seen in rodent mutants have their counterparts in the human species. There are a number of segmented progeroid syndromes where senescence initiates either in utero (leprechaunism, trisomy 21, Mulibrey-nanism), just after birth (xeroderma pigmentosum), after weaning (progeria, Cockayne syndrome, Rothmund syndrome), on or around adrenarche (osteopetrosis), or during young adulthood (Werner syndrome, myotonic dystrophy). In all such conditions where the data is available, immune system deficits and thymic abnormalities are observed (Cardarelli 1989a).

TIN AND THE LIFE PATTERN

The Life Pattern Hypothesis suggests that exogenous tin is synthesized by the thymus into one or more circulating factors that act in the regulation of cell growth, including not only those of the lymphatic system involved in immune function, but also in a wide repertoire of somatic cells. There are at least three possible functions - maturation of specific elements of the immune system, homeostasis in post-mitotic cells, and regulation of mitotic rate in dividing cells.

Since altricial animals are born lacking essential immune function due to immaturity of T and B lymphocytes, and tin is postulated tin as being involved in maturation processes, then one might suspect that tin is absent or in deficient quantities before birth, but present after birth. How does this square with known data? In our studies, we found that the fetal mouse is devoid of tin - or the concentration is below our practical detectability threshold of 2 ppb, that maternal colostrum and milk is high in tin, and the neonate rapidly gains tissue tin content during the nursing period (Cardarelli 1986a). The normal mouse reaches around 12 ppm thymus tin content and complete immune function just prior to weaning. It is notable that in the leukemia prone AKR strain, thymus tin ranges from below detection in 86 of 100 yound adult mice to an average 2.1 ppm in the other 14 (Cardarelli et al 1984b). In general, tin does not cross the placential barrier in mammal, bird, amphibian, reptile, or fish. Human fetuses and stillborns are likewise nearly devoid of tin, but maternal lactates and also bovine milk contains it (Dingle et al 1938; Schroeder et al 1964; Von Mallinkrodt et al 1969). At birth, infant tissue tin is detectable only in the lungs and may thus have entered with the first breath.

Xenobiotic tin homes in on the thymus, and if the amount administered surpasses the processing capability of the thymus, then toxic effects can be observed. Since immune capacity of the female mammal (man and rodent), and the respective thymic are larger, then one would expect a gender difference in toxicity to occur. A recent report indicates that at least with one organotin, this is indeed the case (Sherman et al 1988).

If the thymic tin factors were acting as a hormone or chalone involved in growth regulation, presence of tin in the cell nucleus would be expected. In the only known cell fractionation studies performed, the majority of labelled tin was found in the nuclear fraction (Evans et al 1979, 1980). Specific organotins are known to inhibit DNA, RNA, and protein synthesis which are nuclear processes.

An increase in tin intake results in prolongation of life span which likely arises from extended immune vigor. Consequently enhanced dietary tin should retard thymus involution. In preliminary studies, there is some indication that this assumption is correct (Sherman et al 1986).

The tin factor might be considered as a chalone since it appears to repress gene expression responsible for profligate division of cancer cells and/or other essential segments of genome. In either case, the end result is cell death. In my own observation, cell death does not conform to the characteristics of necrosis, but rather appears to be apoptotic (programmed) such as seen in radiation induced thymocyte interphase death (Borisova et al 1987; Kerr et al 1972; Yamada et al 1988). Observation under the light microscope indicates blistering of the cell membrane and fragmentation of the nucleus, similar to that described for apoptosis, and similar to steroid-induced lethal changes (Wyllie et al 1982). Cytotoxic T lymphocyte destruction of cancer cells follows a fimilar pattern (Russell et al 1982). Electron microscopy analysis of thymic cell death induced by organotin should also demonstrate the apoptotic schema.

The thymus-pineal axis as mentioned supra is involved in sexual maturation in many mammalia, the pineal inhibiting and the thymus accelerating gonadal changes. Is tin somehow involved in the programmed alterations of homeostasis necessary to achieve gonadarche? One might theorize that puberty occurs when thymus derived chalone(s) depress pineal function. The literature is mostly silent though low level trialkyltin exposure is known to cause a superimposition of male characteristics on femal Nassarius (Smith 1981).

Let us consider the question by analogy. The mammalian thymus-pineal axis has its counterpart in the allatum-prothoracic gland axis of insecta (Wiener 1968). The corpora allata secretes juvenile hormone which prevents metamorphosis and the prothoracic gland releases ecdysone which stimulates it. (Mammalian puberty is essentially the equivalent of metamorphosis in insecta). It is noted that low dosages of pesticidal organotins inhibit normal instar molting, pupation, and adult emergence in the mosquito (Cardarelli 1985). Similar results are seen with juvenile hormone. Prevention of insect metamorphosis by environmental manipulation will extend lifespan in the juvenile state by factors of 6X or more (Laufer 1982). Exposure to juvenile hormone similarly prevents adulthood. It has been noted that tin enhanced diets increase rodent lifespan and also retard puberty. Since the corpora allatum fulfills a pineal-like function, might it contain tin? Cholesteryl tri-n-butyltin adipate, one of the anticancer steroids showed a degree of juvenile hormone-like activity against several insect species (Sharma et al 1988). It was concluded that the tin moiety may be associated with growth, hormonal and/or developmental processes of insects. (As an aside, it might be of interest to note that insect juvenile hormone is found in the human thymus!)

Recently I and my colleagus removed a bovine pineal and analyzed it for a number of trace metals by atomic absorption spectroscopy. The results of this initial effort are shown below: (Wet weight basis, average of duplicates)

Calcium	230	ppm
Magnesium	170	ppm
Iron	150	ppm
Nickel	58	ppm
Zinc	28	ppm
Chromium	16	ppm
Tin	14	ppm
Manganese	8	ppm
Copper	5.5	ppm

Whether the similarity between bovine pineal tin (14 ppm) and bovine thymus tin content (11 ppm) has any significance is unknown.

The structure of the natural thymic tin factor has never been ascertained, but is presumed to be a steroid (Cardarelli 1988, Potop 1970). In the recent analysis of the thymi of four mixed breed dogs, the chloroform extractable (lipid) fraction of whole thymus contained 23.8 ppm tin whereas the non-lipid residue, basically protein, had but 6.2 ppm (Sherman et al 1989). It can be presumed that the lipid-tin is a steroid derivative, but what of the non-chloroform extractable material? Are there also tin peptides involved? The work with the Potop et al fractions showed that a number of basically protein isolates contained significant tin

(Cardarelli et al 1988, Potop et al 1970).

An attempt to relate thymus tin content to animal lifespan was not successful. Man, cow, and mouse thymi contain ca. 12 ppm; however, the rat, with 42 ppm and dog with 30 ppm negate the premise (Sherman et al 1989). It is noted that thymic tin content probably increases with age since storage is in the medulla, and involution affects the cortex to a greater extent, and thus some sort of strict comparion based on age is necessary. It is of interest that chelonian species, the only animal with confirmable maximum life span in excess of man's 112 years, have very high thymic tin. In our analysis of four 5-year-old Chelonia mydas, the green sea turtle, an average tin concentration of 456 ppm (dry weight) was found (Cardarelli 1989). This was significantly higher than iron, copper, zinc, chromium, and magnesium content - all known vital turtle nutrients. Also, two land tortoises showed 480 ppm tin based on dry thymus weight.

SUMMARY

It is generally recognized that mammalian senescence is a function of immunoendocrine decline. The thymus gland controls the in utero and neonatal development of the immune system in man and other higher animals. Such development is mediated in great part by thymic humoral factors.

It has been postulated that one or several such factors are tin-bearing hormones (or chalones) that act at the genome level in recipient cells through alteration of histone switches. Both direct and indirect evidence supports this contention. Tin, in organic or inorganic form, is known to accumulate in the thymus, is seemingly processed there into an anticarcinogenic factor that enters lymphatic circulation. Action at the target cell seems to be in the genome. Elevated dietary tin extends rodent life span while depressing incidence of malignancy. Tin deprivation leads to cachexia and a symptomatology closely resembling that seen after neonatal thymectomy of altricial mammals.

Tin content has been found to be extremely high in the thymus of long-lived turtle species.

In utero and neonatally, during human ontogeny, both the immune and developmental function of the thymus are programmed and regulated by higher centers through a distinct thymus-pineal axis. The circadian and circannual rhythmicity of immune responsiveness relates to the transductive effect of the pineal on neural signals received from the hypothalamus. It is hypothesized that a concilience of immunoendocrine function and pineal modulated biorhythms that control major life events - such as parturition, adrenarche, puberty, climacteric, and death, lies in known hypothalamic oscillators timed by the periodicity of environmental electromagnetic radiation. The possible role of the tin hormones in the "Life Pattern" hypothesis of aging is explored based upon very sparse data. Evidence suggests that 1) tin hormones or chalones of thymic origin exist, 2) that they act on the genome in such a way as to 3) destroy malignant cells, and 4) to retard the onset of senescence. An argument is developed from analogy with organotin retardation of insect metamorphosis and slowing of mammalian maturation in both cases being based upon a mediating endocrine axis.

REFERENCES

Anisimov VN, Khavinson VKh, Morozov VG (1982) Carcinogenesis and aging. IV. Effect of low-molecular-weight factors of thymus, pineal gland and anterior hypothalamus on immunity, tumor incidence and lifespan of C3H/Sn mice. Mech Ageing Dev. 19:245-258

Baarlen JV, Schurman HJ, Huber J (1988) Acute thymic involution in infancy and childhood: A reliable marker for duration of acute illness. Hum Path 19:1155-1160

Baertschi AJ, Beny JL, Gahwiler B (1982) Hypothalamic paraventricular nucleus is a privileged site for brain-pituitary interactions in a long-term tissue culture. Nature 295:145-147

Barath P, Csaba G (1974) Histological changes in the lung, thymus and adrenal in two years after pinealectomy. Acta Biol Acad Sci Hung 25:123-125

Bartke A (1964) Histology of the anterior hypophysis, thyroid and gonads of two types of dwarf mice. Anat Rec 149:225-236

Besedovsky H, Sorkin E (1974) Thymus involvement in female sexual maturation. Nature 249:356-358

Bilgicer KI, Sherman LR, Cardarelli NF (1985) Analysis of tin in mice and human organs. J Nutr Gwth Cancer 2:107-115

Bischoff F, Bryson G (1976) Toxicological studies of tin needles at the intrathoracic site of mice. Res Commun Path Pharm 15:331-340

Blask DE (1984) The pineal: An oncostatic gland? In: Reiter RJ (ed) The pineal gland. Raven Press, New York, p 253

Borisova EA, Chukhlovin AB, Seiliev AA, Zherbin AA, Zhivotovsky BD, Hanson KP (1987) Degree of chromatin fragmentation and frequency of nuclear pyknosis in Percoll-fractionated thymocytes of irradiated rats. Internat J Rad Biol 51:421-428

Brainard GC, Rollag MD, Northrup BA, Cole C, Barker FM (1986) The influence of light irradience on pineal physiology of mammals. In: O'Brien PJ, Klein DC (eds) Pineal and retinal relationships, Academic Press, New York p253

Cardarelli NF (1985) Controlled release organotins. 2. Plastic matrices. Rev Si Ge Sn Pb Cmpds 8:313-326

Cardarelli NF (1986a) Experimental studies of tin in the mouse. In: Cardarelli NF (ed) Tin as a vital nutrient. CRC Press, Boca Raton, FL p49

Cardarelli NF (1986b) Tin and the thymus gland. Si Ge Sn Pb Cmpds 9:307-321

Cardarelli NF (1986c) Tin and cancer: A review. In: Cardarelli NF (ed) Tin as a vital nutrient. CRC Press, Boca Raton, FL p35

Cardarelli NF (1986d) Tin and the aging process. In: Cardarelli NF (ed) Tin as a vital nutrient. CRC Press, Boca Raton, FL p235

Cardarelli NF (1986e) Thymic hormones. In: Cardarelli NF (ed) Tin as a vital nutrient. CRC Press, Boca Raton, FL p133

Cardarelli NF (1986f) The thymus gland. In: Cardarelli NF (ed) Tin as a vital nutrient. CRC Press, Boca Raton, FL p99

Cardarelli NF (1988a) Tin as a vital nutrient: Environmental ubiquity. In: Zuckerman JJ (ed) Tin and malignant cell growth. CRC Press, Boca Raton, FL p1

Cardarelli NF (1988b) Tin steroids as anticancer agents. In: Zuckerman JJ (ed) Tin and malignant cell growth. CRC Press, Boca Raton, FL p53

Cardarelli NF (1989a) The thymus in health and senescence, Vol. II. CRC Press, Boca Raton, FL

Cardarelli NF (1989b) The thymus in health and senescence, Vol. I. CRC Press, Boca Raton, FL

Cardarelli NF (1989c, to be published) Tin and the thymus gland: A review.

Cardarelli NF (1990, to be published) Exogenous control of senescence.

Cardarelli NF, Quitter B, Allen A, Dobbins E, Libby E, Hager P. Sherman LR (1984a) Organotin implications in anticarcinogenesis: Background and thymus involvement. Austral J Exp Biol Med Sci 62:199-208

Cardarelli NF, Cardarelli B, Marioneaux M (1984b) Tin as a vital nutrient. J Nutr Gwth Cancer 1:181-194

Cardarelli NF, Cardarelli B, Libby E, Dobbins E (1984c) Organotin implications in anticarcinogenesis: Effects of several organotins on tumor growth rate in mice. Austral J Exp Biol Med Sci 62:209-214

Cardarelli NF, Cardarelli BM (1985e) Method and composition for the detection of a precancerous and leukemic condition in mammals. United States Patent 4511551

Cardarelli NF, Kanakkanatt SV (1985b) Tin steroids and their use as antineoplastic agents. United States Patent 4541956

Cardarelli NF, Fazely F (1988) The correspondence between tin and antiproliferative activity in a series of thymic extracts. Thymus 11:123-126

Carrara M, Zampiron S, Cima L, Sindeplari L, Trincia L, Voltarel G. (1988) Inhibitory properties of tin (IV) diethyldithiocarbamates on tumoral cell growth. Pharmacol Res Commun 20:611-612

Cockerill O, Chang LW, Hough A, Bivens F (1987) Effects of trimethyltin on the mouse hippocampus and adrenal cortex. J Toxicol Environ Hlth 22:149-161

Cremer-Bartles G, Krause K, Mitoskas G, Brodersen D (1984) Magnetic field of the earth as an additional zeitgeber for endogenous rhythms. Naturwiss 71:567-574

Crowe AJ, Smith PJ, Atassi G. (1980) Investigation of the antitumor activity of organotin compounds I. Diorganotin dihalide and dipseudohalide complexes. Chem Biol Interact 32:171-178

Damian E (1985) Melatonin-free pineal substance, an equilibrating hormone. Rev Roum Med-Endocrinol 23:91-96

Dingle H, Sheldon JH (1938) Spectrographic examination of the mineral content of human and other milk. Biochem J 32:1078-1086

Duquesnoy RJ, Kalpaktsloglou PK, Good RA (1970) Immunological studies of the Snell-Bagg dwarf mouse. Proc Soc Exp Biol New York 133:201-206

Ernström U, Karlsson P, Soder, O (1988) Isolation of a thymocyte growth peptide from human thymus. Arch Allerg Appl Immunol 85:434-440

Escomel E (1939) La plus jeune mére du monde Presse Med 47:744-745

Evans WH, Cardarelli NF, Smith DJ (1979) Accumulation and excretion of ^{14}C bis(tri-n-butyltin) oxide in mice. Toxicol Environ Hlth 5:871-877

Evans WH, Cardarelli NF (1980) Chemodynamics and environmental toxicology of controlled release organotin molluscicides. In: Baker R (ed) Controlled Release of Bioactive Materials. Academic Press, New York, p357

Galkin BM, Feig SA, Patchefsky AS, Muir HD (1987) Elemental analysis of breast calcifications. In: Brunner S, Longfeldt B (eds) Breast cancer: Present perspectives of early diagnosis. Springer, Berlin p89

Gielen M (1986) Antitumor active organotin compounds. In: Cardarelli NF (ed) Tin as a vital nutrient. CRC Press, Boca Raton, FL p169

Goidl EA (1987) Aging and immune response Marcel Dekker, NY

Goldman DB, Darrow JM (1983), The pineal gland and mammalian photoperiodism. Prog Neuroendocrinol 37:386-396

Gould SJ (1976) Human babies as embryos. Nat Hist 85:22-26

Guta-Socaciu C, Giurgea R, Rosioru C (1986) Modification of thymus and bursa fabricii induced by N-butyltin pesticide treatment in chickens. Agressol 27:123

Gwinner E (1981) Annual rhythms: Perspective. In: Aschoff J (ed) Handbook of behavioral neurobiology, Vol IV. Plenum, NY p381

Hartge R., Jenkins DM, Kohler HG (1987) Low thymic weight in small-for-dates babies. Eur J Obstet Gynec Reprod Biol 8:153-155

Hollwich F, Dieckhues B (1974) Changes in the circadian period of blind people. In: Scheving LE, Hallberg F, Pauly JE (eds) Chronobiology. Igaku Shoin, Tokyo p285

Inouye ST, Kawamura H (1979) Persistence of circadian rhythmicity in a mammalian hypothalamic "island" containing the suprachiasmatic nucleus. Proc Nat Acad Sci 76:5962-66

Kendall MD (1981) The thymus gland. Academic Press, New York

Kerr JFR, Wyllie AH Currie AR (1972) Apoptosis: A basic biological phenomenon with wide ranging implications in tissue kinetics. Br J Cancer 26:239-257

Kurumado K, Mori W (1976) Synaptic ribbons in the human pinealocyte. Acta Path Jpn 26:381-384

Lapin V (1976) Pineal gland and malignancy. Öst Z Onkol 3:51-60

Laufer H (1982) Aging in insects. In: Regelson W, Sinex FM (eds) Intervention in the aging process B: Basic research and preclinical screening. Alan Liss, New York p307

Leng G, Mason WT, Dyer RG (1982) The supraoptic nucleus as an osmoreceptor. Neuroendocrinol 34:75

Levine S (1986) The effects of metallic tin and inorganic tin on plasma cells. In: Cardarelli NF (ed) Tin as a vital nutrient. CRC Press, Boca Raton, Florida p95

Levine S, Saad A, Rappaport I (1987) The effect of metallic tin on background plaque-forming cells, immunoglobulin and the immune response. Immunol Invest 16:201-212

Levine S, Sowinski R (1982) Plasmacellular lymphadenopathy produced in rats by tin. Exp Molec Pathol 36:86-98

Luckey TD (1973) Thymic hormones. University Park Press, Baltimore

Maestroni GJM, Conti A, Pierpaoli W (1988) Pineal melatonin, the fundamental immunoregulatory role in aging and cancer. Ann New York Acad Sci 521, 521:140-148

Magee K, Basinska J, Quarrington B, Stancer HC (1970) Blindness and menarche. Life Sci 9:7-12

Michaelson SM (1986) Interaction of nonmodulated radiofrequency fields with living matter: Experimental results. in: Polk C, Postow E (eds) Handbook of biological effects of electromagnetic fields. CRC Press, Boca Raton, Florida p339

Miles LE, Raynal DM, Wilson MA (1977) Blindman living in normal society has circadian rhythms of 24.9 hours. Science 198:421-423

Miller JFAP (1964) Effect of thymic ablation and replacement. In: Good RA, Gabrielsen AE (eds) The thymus in immunobiology. Harper and Row, New York p436

Miller K, Scott MP, Foster JR (1984) Thymic involution in rats given diets containing dioctyltin dichloride. Clin Immunol Immunopath 30:62-70

Møller M (1978) Presence of a pineal nerve (nervus pinealis) in the human fetus: a light and electron microscopic study of the innervation of the pineal gland. Brain Res 154:1-12

Møller M (1979) Presence of a pineal nerve (Nervus pinealis) in fetal mammals. In: Kappers JA, Pevet P (eds) The pineal gland in vertibrates including man. Elsevier/North- Holland, Amsterdam New York p. 103

Osipov AI, Desyatov VP (1971) The mechanism of the influence of solar activity oscillations on the human organism. In: Gnevyshev MN, Oi AI (eds) Effects of solar activity on earths atmosphere and biosphere. Akad Nauk SSSR, Moskva p232

Pengelley ET (1974) Circannual Clocks: Animal Biological Rhythms, Academic Press, New York

Penninks AH, Punt PM, Bol-Schoemakers M, van Rooijen HJM, Seinen W, (1986) Aspects of the immunotoxicity, antitumor activity and cytotoxicity of di- and trisubstituted organotin halides. Si Ge Sn Pb Compds 9:367-380

Peterson RF, Cardarelli NF (1986) Aspects of xenobiotic tin distribution in rodent tissue. In: Cardarelli NF (ed) Tin as a vital nutrient. CRC Press, Boca Raton, Florida p59

Potop I, Sterescu V, Boru V, Petoni R, Petrescu E, Ghinea E (1970) Effect of an "S" purified thymus factor (isolated from IIB3 thymus fraction) on the in vitro proliferation of tumor cells. Neoplasma 17:655-660

Reiter RJ (1981) Chronobiological aspects of the mammalian pineal gland. Prog Clin Biol Res 59:223-233

Robertson AJ (1964) The romance of tin. Lancet 1:1229-36

Russell VH, Massakowski V, Rucinsky T, Philips G (1982) Mechanisms of immune lysis. III. Characterization of the nature and kinetics of the cytotoxic T lymphocyte-induced nuclear lesion in the target. Immunol 128:2087-94

Schroeder HA, Balassa JJ, Tipton IH (1964) Abnormal trace metals in man. J Chron Dis 17:483-502

Schroeder HA, Balassa JJ (1967) Arsenic, germanium, tin and vanadium in mice: effects on growth, survival and tissue levels. J Nutr 92:245-262

Schwarz K (1974) Recent trace element research exemplified by Tin, Fluorine and silicone. Fed Proc 33:1748-57

Schwarz K, Milne DB, Vinyard E (1970) Growth effects of tin compounds in rats maintained in a trace element controlled environment. Biochem Biophys Research Commun 40:22-29

Seinen W, Willems MI (1976) Toxicity of organotin compounds. I. Atrophy of thymus and thymus-dependent lymphoid tissues in rats fed di-n-octyltin dichloride. Toxicol Appl Pharmacol 35:63-75

Sharma R, Tare V, Bhonde SB (1988) Bioactivity of some organotin compounds on insects. In: Zuckerman JJ (ed) Tin and malignant cell growth. CRC Press, Boca Raton, Florida p201

Sherman L (1986) Tin, tumors and the thymus gland. In: Cardarelli NF (ed) Tin as a vital nutrient. CRC Press, Boca Raton, Florida p71

Sherman LR, Masters JG, Peterson R, Levine S (1986) Tin concentration in the thymus glands of rats and mice and its relation to the involution of the gland. J Anal Toxicol 10:6-9

Sherman LR, Koch JR, Vinton DS (1988) Gender related toxic effects of tri-n-butyltin taurocholate and tri-n-butyltin glycocholate in rats. J Pennsylvania Acad Sci 62:152-154

Sherman LR, Cardarelli NF (to be published 1989) Tin in the thymus glands of dogs. Thymus 12:

Smith BS (1981) Tributyltin compounds induce male characteristics on female mud snails Nassarius obsoletus. J Appl Toxicol 1:141-144

Snoiej NJ, Bol-Schoenmakers M, Penninks AH, Seinen W (1988) Differential effects of tri-n-butyltin chloride on macromolecular synthesis and ATP levels of rat thymocyte subpopulations obtained by centrifugal elutriation, Int J Immunopharmacol 10:29-37

Tapp E (1980) The human pineal gland in malignancy, J Neural Transm 48:119-129

Uede T, Ishii Y, Matsume A, Shimagawara J, Kikuchi K (1981) Immunohistochemical study of lymphocytes in rat pineal: Selective accumulation of T lymphocytes. Anat Rec 199:239-247

Vollrath L (1986) The postnatal differentiation of the mammalian pineal complex. In: Gupta D, Reiter RJ (eds) The Pineal Gland During Development: From Fetus to Adult, Croom Helm, London p56

Von Mallinckrodt G, Pooth M (1969) Simultaneous determination of twenty-five metals and metalloids in biological materials, Arch Toxicol 25:5-30

Waldhauser F, Gisinger B (1986) The pineal gland and its development in human puberty. In: Gupta D, Reiter RJ (eds) The pineal gland: From fetus to adult. Croom Helm, London 1986 p134

Walford RL (1969) The immunologic theory of aging, Williams and Wilkins, Baltimore

Walford RL (1983) Maximum life span. Avon, New York

Wiener H (1968) External chemical messengers. V. More functions of the pineal gland, New York State J Med 68:1019-38

Wyllie AH, Morris RG (1982) Hormone-induced cell death. Purification and properties of thymocytes undergoing apoptosis after glucocorticoid treatment. Amer. J. Pathology 109:78-87

Yamada T, Ohyama H (1988) Radiation-induced interphase death of rat thymocytes is internally programmed (apoptosis) Int J Rad Biol 53:65-75

Zacharias L, Wurtman R (1964) Blindness: Its relation to age of menarche. Science 144:1154-55

Zeppezauer M, Reichhart R, Jornvall H (1988) Homeostatic thymic hormone: Chemical properties and biological action. In: Zuckerman JJ (ed) Tin and malignant cell growth. CRC Press, Boca Raton, Florida p155

THE SPECIATION AND BIOAVAILABILITY OF TIN IN BIOFLUIDS

J.R.Duffield, C.R.Morris, D.M.Morrish, J.A.Vesey and
D.R.Williams
School of Chemistry and Applied Chemistry, University
of Wales, Cardiff, United Kingdom

1. INTRODUCTION

1.1 Historical Outline

Tin has been mined and utilized from antiquity. The ancient metallurgists called it diabolus metallorum, the devil among metals, because its presence could make an alloy hard and brittle [1] and, to this day, we describe something to its detriment as being "tinny". This so called "Cinderella" metal [2] finds application in plastics, alloys, solders, pharmaceuticals, etc (the world resources may last less than 50 years [3])but its most obvious role in modern society is in the production of tinplate for food preservation by canning. The initial discovery that decay of foodstuffs could be halted by heating them in sealed glass vessels was made in 1809 by Nicolas Appert, a Parisian confectioner and, the following year, an Englishman Peter Durand patented the idea of using "glass, pottery (or) tin" (tinplate) with Appert's technique. The phenomenal success of this discovery is attested to by the fact that tin, a naturally non-ubiquitous element, is now present in small amounts in practically all animal life.

While soils contain up to 200 ppm tin, the ocean contains only 3 ppb and air contains between 0.002 ppb in rural areas and 2.3 ppb in industrial centres [3]. With an average concentration of 3 ppm in the lithosphere, tin occurs with approximately the same abundance as cobalt and is much rarer than zinc, copper or lead. Most noticeably, it is isolated in discrete regions of the earth's crust, and vast areas of the northern latitudes apparently lack tin deposits. It has been said that for an element to perform a necessary biological function, it must be uniformly available or living organisms would be localized in areas where it was present. Since tin was not universally available before the nineteenth century, it has been tentatively suggested that it was not an essential trace element for plant life in general or, by extension, for animal life [4]. Tin is reported to be a micronutrient in the rat [5] but no deficiency state is recognised in man and the fact that an average 70 kg man has been calculated to contain 0.35g of the metal [6] may indicate only that tin is a fairly innocuous industrial contaminant.

Of the average tin intake of 2-3mg/day in the UK in 1984, canned food which formed 6-7% by weight of the normal person's diet, contributed by far the greatest percentage [7]. In 1908 the maximum permitted level of tin in canned foods in the UK was fixed at 286 ppm [8]. This figure was lowered to 250 ppm in 1953 "since a high tin content in food would be contrary to good commercial practice" [8] and a recommendation to reduce it still further to 200 ppm was made in 1983 [9].

Tin has also found major applications in the pharmaceutical industry. Since Brown's initial finding in 1972 that tumour growth in mice was retarded by triphenyltin acetate [10], several organotin compounds have demonstrated a similar activity and the use of tin in tumour therapy is well documented [11].

A traditional belief that the tin workers of Beauve never suffered from furunculosis led to the use of tin and its compounds in the treatment of staphylococcal infections. In 1917, Frouin and Gregoire [12-14] published a study which suggested that tin, tin oxide, stannous chloride and sodium stannate modified the virulence of staphylococci. This work, based on seven infected rabbits and two control animals, has since been discredited and more recent evidence indicates that neither soluble nor insoluble compounds of tin have any effect on staphylococci either in vitro or in vivo [15,16]. Nevertheless, the belief that tin had some value in the control of cutaneous sepsis persisted well into this century and resulted in an occurrence of organotin poisoning in France in 1954. The proprietary compound involved, "Stalinon", had become contaminated with triethyltin impurities and this resulted in the deaths of more than 100 people [17]. In the UK, a preparation containing methyl stannic iodide "Staniform" [18] was available before 1958 as an external treatment for staphylococcal infections and an oral treatment containing tin powder and tin(II) oxide "Stannoxyl" was available until recently [11]. In contrast to the organotins, metallic tin and inorganic compounds are generally non-toxic [19] with the notable exception of stannane, SnH_4, which is more toxic than arsine [20].

For many years, stannous compounds have been used in dentistry and oral hygiene. As long ago as 1947, it was known that tin(II) fluoride protected dental enamel from dissolution in lactic acid [21] and since then SnF_2 has been incorporated into dentifrices, mouthwashes, topical solutions and occasionally, dental cements. The compound appears to exert its prophylactic and therapeutic effects in several ways. In combination with acidulated phosphofluoride it allows control of dental caries; it inhibits dental plaque growth; it effectively controls root hypersensitivity; it reduces root surface solubility and finally, it causes less mottling than sodium fluoride [11].

It has been said that the clinical effect of antimicrobial agents as plaque growth suppressors is not well correlated with in vitro antimicrobial assays of the same agents [22]. Reports indicate that while both stannous chloride and stannous fluoride possess bacteriostatic effects on oral micro-organisms in vitro [23], growth inhibition by $SnCl_2$ in vivo is only slight [24]. However, no explanation has been offered for these different effects and the results may prove to be anomalies caused by the experimental difficulty in keeping stannous ions in solution. It has been shown that commercial toothpastes containing either SnF_2 alone or in combination with tin(II) pyrophosphate, $Sn_2P_2O_7$ [25], function as efficient plaque inhibitors and it appears that the tin(II) ion is the bacteriostatic agent [26] with little, if any, effect from the fluoride component. Most of the tin retained in the mouth is bound to the epithelial surfaces, the dental plaque and salivary macromolecules and is slowly released over a period of up to four hours following brushing, thus allowing a long lasting bacteriostatic effect [25]

1.2 Speciation Studies

It has long been realised that the bioavailability and toxicology of a trace element is related not just to the absolute concentration of that element, but also to its valence state, complexation, solubility, etc. Thus, the precise chemical form in which an element occurs, i.e. its speciation, has the most profound effect on its absorption, distribution and excretion in vivo.

The direct study of the speciation of a trace element in its biological context is difficult since, by definition, the concentrations of the free element and its complexes are low and they are not readily measured by standard laboratory procedures. In such circumstances, computer modelling can provide a calculated speciation profile based on specified component concentrations and a comprehensive database of relevant formation constants. In recent years, the most abundant low-molecular-weight complexes of Ca^{2+}, Mg^{2+}, Mn^{2+}, Fe^{3+}, Cu^{2+}, Zn^{2+} and Pb^{2+} in human blood plasma have been identified in the presence of both naturally occurring ligands and synthetic chelating agents using computer modelling. The technique has been used to rationalise the efficacy of drugs already prescribed and to predict possible new therapeutic agents [27].

At Cardiff, the MINEQL program [28] has been used for the calculated speciation of such diverse scenarios as essential metals in digested foods, actinide disposal in nuclear waste repositories and precipitates formed in total parenteral nutrition fluids (which are given to patients who cannot be sustained normally). MINEQL requires initial data relating to the ionic charge of components, their free and total concentrations, the ionic strength of solution and, as mentioned, the formation constants of possible complexes. The free concentrations and ionic strength are estimated and then refined by the program which solves a series of mass balance equations and presents the calculations as percentage of an element bound in a given species. The program specifically applies to systems in thermodynamic equilibrium (the steady-state in vivo approximates to this) and it does not take account of either kinetic or sorption phenomena. Thus, the primary role of computer simulation models is in identifying important components and complexes which warrant further experimental study.

The production of a valid speciation profile is obviously dependent upon the accuracy of the formation constants in the database. The stannous complex constants were obtained from three sources:
i) literature values of which there are few (because solutions of tin(II) readily oxidize, hydrolyse and precipitate)
ii) systems characterized in this study i.e. stannous citrate and malate
iii) estimated values based on chemical analogues.

The modelling has been confined to canned fruit juices, (which have been implicated in tin poisoning incidents [29]), and to stannous complexes in dentifrices, in an attempt to identify the most abundant species both before ingestion and in the mouth.

2. EXPERIMENTAL

2.1 Materials

Stannous chloride (BDH); citric acid (BDH AnalaR); L(-)-malic acid (Koch-Light) and all other reagents were used without further purification. The solutions were prepared in distilled, degassed, deionised water, and a background ionic strength of 150 mmol dm^{-3} with respect to chloride ions was maintained throughout by the addition of sodium chloride (BDH AnalaR). Tin(II) solutions were freshly prepared in acid media to minimize hydrolysis and were maintained under deoxygenated nitrogen to limit oxidation. The metal content of these solutions was determined by titration against potassium iodate using amaranth as indicator [30] and the acid content was determined by titration against a standard alkaline solution (sodium hydroxide, BDH ConvoL)

2.2 Procedure

The ligand protonation constants and metal-ligand formation constants were determined by glass electrode potentiometric titration at 37°C as described by Filella and Williams [31]. Various total ligand, proton and metal concentrations were used as detailed in Table 1. The experimental curves; the calculated values of protonation and formation constants and the simulated curves were all obtained using the ESTA (Equilibrium Simulation for Titration Analysis) suite of computer programs [32].

For ligand protonation studies, where the species formed have been previously defined, parameters such as the electrode constant, $E°$, could be refined with the formation constants, in order to reduce the systematic errors in the system. However, for metal-ligand interactions, where the species present have not been unequivocally identified, only the formation constants were refined. Although the hydroxyl

TABLE 1: Summary of Potentiometric Titration Data
a = Initial concentration in mmol dm^{-3} in vessel
b = 17.7 mmol dm^{-3} solution added from burette
c = 18.2 mmol dm^{-3} solution added from burette

System	Titration	Number of Points	a [L]	a [M]	$-\lg[H]$ range
Citrate Proton-	1	101	10.4	–	2.5 – 11.1
	2	59	5.4	–	2.7 – 11.1
	3	133	15.2	–	2.4 – 10.9
	4	127	15.2	–	2.4 – 10.7
Citrate-Sn(II)-Proton	1	85	16.0	3.7	1.6 – 5.9
	2	68	13.3	6.2	1.4 – 4.5
	3	57	15.2	4.7	1.5 – 6.0
	4	57	12.7	3.1	1.7 – 5.2
	5	42	11.4	4.7	1.5 – 5.4
	6	37	7.3	5.4	1.5 – 4.3
	7	49	16.9	3.1	1.7 – 4.6
	8	20	10.1	9.4	1.4 – 4.9
	9	48	10.1	9.2	1.3 – 4.6
	10	51	13.1	2.6	1.8 – 5.6
Malate Protontion	1	47	11.4	–	2.6 – 5.6
	2	68	15.9	–	2.6 – 5.8
	3	38	6.0	–	2.8 – 6.3
	4	51	10.2	–	2.6 – 6.1
	5	71	17.5	–	2.5 – 6.2
Malate-Sn(II)-Proton	1	52	15.1	b	4.1 – 2.3
	2	41	15.1	b	4.0 – 2.5
	3	42	15.1	c	4.0 – 2.5
	4	43	15.1	c	4.1 – 2.5
	5	42	15.1	c	4.0 – 2.5
	6	43	15.1	c	4.1 – 2.5

TABLE 2: Formation Constants at 37°C, I = 150 mmol dm^{-3} Cl^{-*}

$\beta_{pqr} = [M_pL_qH_r]/[M]^p[L]^q[H]^r$

Interaction	Species p q r	lg β_{pqr} (s.d.)	Objective Function	R Factor	No. of Points	No. of Titrations
Citrate-Proton	0 1 1 0 1 2 0 1 3	5.511 (0.003) 9.725 (0.003) 12.512 (0.004)	1034.0	0.010	420	4
Citrate-Sn(II)-Proton	1 1 0 1 2 0 1 1 1 1 2 1 1 1 -1 1 2 2 1 1 -2	7.94 (0.02) 10.90 (0.07) 10.62 (0.03) 15.44 (0.07) 3.16 (0.03) 19.50 (0.08) -3.12 (0.07)	2508.7	0.010	514	10
Malate-Proton	0 1 1 0 1 2	4.608 (0.003) 7.764 (0.007)	65.9	0.003	275	5
Malate-Sn(II)-Proton	1 1 0 1 2 0 1 1 1	4.87 (0.01) 7.51 (0.02) 7.16 (0.04)	308.2	0.005	263	6

* Formation constants were corrected to zero ionic strength for inclusion in the database by means of the Davies equation.

groups of the organic acids can deprotonate at high values of pH, this is generally beyond the detection limit of the glass electrode and thus the citrate and malate were assumed to have three and two dissociable protonsrespectively.

The affinity of tin(II) for the dicarboxylate malate was, as expected, much less than for the tricarboxylate citrate. Hence, the conventional procedure of titrating alkali against an acidic solution of metal plus ligand used for the determination of stannous-citrate formation constants was not suitable for the stannous-malate system. Instead a modified procedure of titrating acidified metal against alkali plus ligand was adopted for the malate, thus ensuring a very low tin concentration at the highest values of pH. The selection of the best model (i.e. most complete description) of the metal-ligand systems was performed on the basis of minimizing the standard deviations, objective functions and R factors together with a visual comparison between experimental and simulated formation curves.

The speciation profiles of tin in canned foods and dentifrices were calculated by the MINEQL program [28] using component concentration data from tabulated sources [33,34].

3. RESULTS

3.1 Formation Constant Measurement

The experimental formation curves of stannous citrate and stannous malate are shown in Figs. 1a and 2a respectively. The protonation constants of citrate and malate, together with the formation constants of their tin(II) complexes are detailed in Table 2. These constants, calculated by the OBJE task of ESTA [32] were used to produce simulated formation curves for comparison with the experimental plots, and the calculated formation curves of the tin(II) interactions with citrate and malate are shown in Figs 1b and 2b respectively.

The protonation constants calculated in this study were in good agreement with previously reported values [35]. For the metal-ligand interactions, only the 1:1 complexes (ML) of the stannous citrate and malate systems have been previously quantified [36,37]. The formation constant obtained by Sherlock and Britton [37] for stannous citrate (lg β ML = 7.7, 30°C, ionic strength not given) is very similar to the figure obtained in this study (7.9) but their value for stannous malate (lg β ML = 7.4) is substantially greater than the figure presented here (4.9). Stannous formation constants are notoriously difficult to measure precisely but well-established trends in suitable chemical analogues (i.e. Fe^{2+}, Co^{2+}, Cu^{2+}, Zn^{2+}, see below) indicate that the formation constant for the malate should be much less than for the citrate (and not of comparable magnitude). Thus, the value for the malate complex presented here may be more accurate than that obtained in previous studies.

TABLE 3: Speciation Profile of Tin in Unsweetened Canned Orange Juice

Figures represent % of total tin bound in a given complex. Only species which contain ⩾3% of total tin are shown. No precipitation calculated.

Sn=Stannous^{2+}; Cit=Citrate^{3-}; Ox=Oxalate^{2-}; Cys=Cysteinate^{2-}; His=Histidinate^{-}; Arg=Arginine^{0}

In Can
pH = 3.7

Complex	Total [Sn(II)] in can in mmol dm^{-3} (ppm)		
	0.42 (50)	2.11 (250)	4.21 (500)
$[SnCit_2H_2]^{2-}$	31.3	31.4	31.2
$[SnCit]^{-}$	26.8	28.2	29.7
$[SnCit_2H]^{3-}$	12.3	12.5	12.5
$[SnOx]^{0}$	10.5	9.9	9.3
$[SnOx_2H]^{-}$	3.8	3.1	2.4
$[SnOx_2]^{2-}$	3.7	3.0	2.4
$[SnCitH]^{0}$	2.7	2.8	3.0

In Mouth
pH = 6.8
(diluted 1:1 by saliva)

Complex	Total [Sn(II)] in can in mmol dm^{-3} (ppm)		
	0.42 (50)	2.11 (250)	4.21 (500)
$[SnCys_2]^{2-}$	53.8	12.0	6.0
$[SnHis_2(OH)_2]^{2-}$	46.1	11.3	5.6
$[SnCit(OH)_2]^{3-}$	–	41.8	54.9
$[SnArg_2(OH)_2]^{0}$	–	20.0	13.9
$[SnCitOH]^{2-}$	–	12.7	16.7

3.2 Regression Analysis

In order to estimate formation constants where experimental data was not available (particularly for tin(II)-amino acid complexes) a technique of regression analysis on selected chemical analogues was used. The charge: crystal ionic radius ratio (Q/r) of the stannous ion has a value (2.15) comparable to Q/r for the divalent cations of cobalt (2.78), copper (2.78), iron (2.70) and zinc (2.70). Plots of known formation constants of stannous complexes against constants of the four individual analogues produced reasonably straight lines whose gradient and intercept could be used to estimate unmeasured stannous formation constants in instances where the formation constant of an analogue was known. The correlation between formation constants of stannous and cupric ions is shown in Fig.3 (some of the common ligands are labelled) and the mathematical relationships between tin(II) and its analogues, as calculated by this technique, are given below.

$$\lg K (Sn^{2+}) = 0.99 \lg K (Co^{2+}) + 2.92$$
$$\lg K (Sn^{2+}) = 0.91 \lg K (Cu^{2+}) + 2.15$$
$$\lg K (Sn^{2+}) = 1.11 \lg K (Fe^{2+}) + 3.14$$
$$\lg K (Sn^{2+}) = 1.06 \lg K (Zn^{2+}) + 2.52$$

No single analogue produced a regression analysis vastly superior to any other, so that the estimated tin constants are generally averages from more than one analogue.

3.3 Speciation Modelling

Tin in Canned Fruit Juice. The speciation profiles of tin in unsweetened orange juice both in the can and mixed (1:1) with saliva in the mouth are given the Table 3. Although the modelling of orange juice only is shown, several other juices were considered and their speciation profiles were very similar, with a predominance of citrate complexes formed in the can and a mixture of citrate and amino-acid complexes formed in the mouth. It was noticeable that, over the range used, (50-500 ppm in the can), the tin concentration has little effect on the distribution between the complexes while they are in the can but it has a marked effect once the juice is diluted by saliva.

For a given metal concentration, whether a complex is formed to any extent is dependent upon three factors: lg K for the complex, the pH, and the ligand concentration. The difference between profiles of a given juice in the can and in the mouth could not be related to the formation constants (which of course are fixed) nor to the change in ligand concentrations (since the metal: ligand concentration remains approximately constant) but instead is the result of a change in pH. The juices considered (orange, grapefruit & tomato) have pH fixed between 3.2 and 4.2 [34] while the pH in the mouth is buffered at about 6.8. If the input files of the juice in the can are run at pH 6.8, the profile produced closely resembles that produced by the juice in the mouth

TABLE 4: Speciation Profile of a Tin-Containing Dentifrice in Water and in Saliva

Only species which contain ⩾3% of total tin are shown.
(s) indicates precipitation of complex.

In Water (1g in 1ml)

Complex	pH			
	4.0	5.0	6.0	7.0
$[SnP_2O_7H]^-$	28.5	6.7	0.8	–
$[Sn(OH)_2]^0$	22.8(s)	48.6(s)	53.5(s)	78.0(s)
$[SnP_2O_7]^{2-}$	16.3	37.5	41.9	18.9
$[SnCit]^-$	16.0	1.2	–	–

In Saliva (1g in 1ml)

Complex	pH			
	4.0	5.0	6.0	7.0
$[SnP_2O_7H]^-$	28.4	6.8	0.8	–
$[Sn(OH)_2]^0$	20.6(s)	47.1(s)	52.9(s)	75.1(s)
$[SnCit]^-$	17.0	1.5	–	–
$[SnP_2O_7]^{2-}$	16.2	37.5	42.1	21.0
$[SnCitOH]^{2-}$	3.1	2.7	1.0	0.3

(ie a marked variation of speciation with tin concentration).
At pH 3.2-4.2, citrate, histidinate, arginine and cysteine are all deprotonated to some extent but, presumably as a result of the high citrate concentrations, complexes of the latter dominate the profile. At higher values of pH in saliva, the probability of forming hydroxyl adducts of histidine and arginine increases and they appear in the calculated models. It should be noted that the stannous complexes of arginine and histidine have not been fully characterised and these hydroxyl adducts are only assumed to form on the basis of chemical analogue behaviour.

The validity of the model is dependent upon a number of assumptions viz. that the database of formation constants at 37°C and zero ionic strength is complete and accurate; that there is no tin(IV) or organotin formation; that precipitation of only stannous pyrophosphate [38] or hydroxide may occur; there is no adsorption to solid phases; that input concentration data is correct and that the juice is completely dissociated into its component amino acids and metal ions. Of these, the assumption that no adsorption to proteins or other solid phases occurs is the least justifiable (see Discussion) but, because the tin precipitated down with proteins is eliminated from the digestive process, the modelling represents worst-case figures (i.e. much less tin will be bioavailable than the calculated speciation suggests). Although the primary purpose of this study is to indicate which complexes merit further investigation, it is encouraging to note that most of the complexes formed in the can and mouth are anionic and not the neutral or cationic species which are reported to be bioavailable (thus complementing the experimental observation that tin salts are very poorly absorbed, see Discussion).

Tin in Dentifrice. The speciation profiles of stannous ions from a tin-containing dentifrice dissolved both in water and in saliva (1g in 1ml) are shown in Table 4. The interactions between tin(II) and some common components of toothpastes (e.g. sorbitol, lauryl sulphate) have not been quantified and their effect upon the speciation is therefore unknown. Since dentifrice formulations which contain tin generally have a lower pH than those which do not, the profiles were produced in the pH range 4.0-7.0

It is immediately apparent from Table 4 that the profiles were practically identical in the two fluids and this was not altogether surprising in view of the fact that ligands from the dentifrice were in far higher concentration than the ligands present in saliva. The precipitation of stannous hydroxide at all pH valves illustrates the difficulty in keeping the stannous component of toothpastes in solution [21]. The precise nature of the binding of tin(II) to the oral mucosa and to the bacteria whose growth it inhibits is unclear. The interaction between stannous ions and bacteria is pH dependent indicating an electrostatic binding to negatively charged surface groups on the bacteria [39] (large polyanions such as glycosaminoglycans may provide similar binding sites on the oral mucosa). On binding to the bacterial surface, the tin may prevent their adhesion to teeth (zinc is reported to have a similar mode of action) [40]. Zinc and tin also inhibit acid production by bacteria and it is likely that they interfere with the metabolic activity of the cells [40].

The speciation studies indicate that much of the tin would be bound in hydroxyl and pyrophosphate complexes and it is likely that the control of dental caries by tin-containing dentifrices is related to an interaction between stannous ions and the components of enamel apatite, $Ca_{10}(PO_4)_6(OH)_2$ [41].

4. DISCUSSION

4.1 Tin in Canned Foods

If one considers the standard redox potentials of the elements as shown in the electrochemical series, the use of zinc as a coating in galvanized iron can be readily understood, because the zinc is preferentially oxidized. For sheet tinplate in air however, the use of tin as a protective coating for the iron is purely mechanical and relies on a complete covering of the underlying metal. Once the tin coating is scratched or broken, the exposed iron is preferentially oxidized and corrodes rapidly (the rusting of scratches on the outside of tin cans is evidence of this).

The tinplate generally employed for canning is not a simple sandwich of steel in tin but rather a multilayer structure with a tin-iron alloy layer between the tin and the steel. The tin coating of these so-called "plain" cans, applied by electrodeposition or hot-dipping, is never continuous and contains a number of cracks and pores which extend through both the tin and alloy layers. Thus, the iron is exposed in part and, on the basis of electrode potentials, corrosion of the base metal would be expected. That, in practice, the iron is not corroded is related to the ready complexation of stannous ions by the organic ligands of the foodstuff. This lowers the potential of the tin half-cell (according to the Nernst expression) so that the tin behaves as a sacrificial anode and is preferentially oxidized.

The corrosion of tin involves the solution of bivalent tin ions which remain in the stannous state [42,43] and it is dependent upon a number of physical and chemical factors [44-46]. It is proportional to storage time, temperature, oxygen in the can headspace and inversely proportional to pH [47]. The relationship between corrosion and oxygen is interesting: it has been observed that at a given pH and temperature, there is a rapid increase in tin concentration for the first three weeks of storage, followed by a more gradual temperature-dependent increase during the next twelve months. Since it is known that oxygen levels are rapidly depleted within the first 100 hours of storage [48], it is likely that oxygen dissolved in the juice or trapped in the can head-space reacts rapidly with tin metal to form stannous ions followed by a second slow anaerobic phase of corrosion (thus lending chemical support to the old adage that food should not be stored in open cans) [49].

The speciation, concentration and corrosion of the tin are all affected by the components of the canned food. For a given ligand concentration, the more stable the complex formed, the greater the expected depression of the potential of the tin and the higher the expected corrosion rate (of the tin) [37]. It has been demonstrated that if the citrate content of canned pears is low and malate is high, tin offers insufficient protection to steel and pitting of the steel occurs [36]. This is indirect evidence that the formation constants of stannous-citrate complexes are much greater than those of stannous-malate complexes (see Results).

In solids [50], fruits and juices [51] much of the tin corroded from the can interior wall is in an insoluble form, bound to proteins and other highly porous solid phases [51]. This is not a true adsoption since, if the complex is transferred to an aqueous solution containing no tin, it does not lose tin to the aqueous phase (i.e. it is not reversible) and has more accurately been described as a "pseudo" adsorption [51]. The insoluble tin-protein complex is not broken up to any extent by either gastric or trypic artificial digestions [52] and the metal combined in this manner is unlikely to exert any direct toxic effect. It may however, exert an indirectly deleterious effect by interfering with the normal digestion of proteins [52] and the possible complicating factor of bacterial action in the gut has not been studied [53].

A considerable proportion of the tin which is not bound to proteins or carbohydrates is said to be in the form of colloids or finely divided insoluble complexes [54]. Thus, only a small percentage of the total tin would be in the form of soluble low- molecular-weight complexes suitable for absorption from the intestinal tract.

It has been shown that in fruit pulp, the anionic fraction, consisting of organic acids, contributes to the corrosion of tinplate but the amino acid and sugar fractions have no significant effect [55], (presumably because the latter do not strongly complex stannous ions). Although Gruenwedel et al have established that sulphur-containing amino acids strongly bind the tin(II), [56,57] very little is known of either the formation constants or the stoichiometry of other stannous - amino acid complexes [58-60].

The sequestering of tin from the can wall is complicated by the fact that some food components accelerate corrosion, (the "cathode depolarizers"), while others (e.g. agar and gelatin [53] act as inhibitors. The cathode depolarizers include oxygen, sulphides and sulphites [53], nitrates (which commonly contaminate tinned tomatoes) [9], and copper [61]. These may be natural constituents of the food or of the water used in preparation. Alternatively, they may be the result of cultivation or processing procedures (e.g. nitrates may be formed from nitrogenous fertilizers [50].)

If it is necessary to maintain a very low level of tin in the canned product, e.g. for infant foods, or if the food contains high levels of cathode depolarizers which are particularly aggressive to tin, lacquered cans are used. In such cans, the whole of the tin surface is covered by the lacquer and corrosion is concentrated on any steel exposed by discontinuities in the varnish. Moreover, the ability of the tin to protect iron at its own expense is hindered and the resultant formation of pinholes generally reduces the shelf life of food in lacquered cans.

It is interesting to note that a total absence of tin in heavily lacquered cans may cause a foodstuff to assume an unacceptable taste and colour. It has been suggested that approximately 20 ppm tin is a necessary component of some canned products [62], and this is said to be due to the reducing action of metallic tin, which aids the preservation of ascorbic acid [43].

Although it is indisputable that very large doses of tin (>1300 ppm) from canned foods have caused acute gastrointestinal disorders [63-65], there is no evidence that toxicity is due to absorption of tin and the most likely cause is local irritation of the alimentary tract [63]. Many of the studies purporting to demonstrate the toxicity have relied on intravenous and subcutaneous introduction of the metal or its salts [20,66] and it is unreasonable to extrapolate such observations to oral administration. In addition, the toxicity of stannous tartrate or stannous citrate solutions given by mouth [42,43] probably bears little relation to the toxicity of tin as it exists in canned foods.

4.2 Bioavailability

Benoy et al [63] have suggested that a tolerance to high levels of tin in the diet may be acquired so that a fruit juice containing 1370 ppm tin caused gastrointestinal disturbance in all (five) volunteers initally but a repeat experiment caused symptoms in only one of the five. This may be related to the observation that there is a decrease in absorption with prolonged exposure [67] and is in contrast to a much earlier study which tentatively suggested that continued feeding of tin salts may cause changes in the gut mucosa which favour absorption of the metal [68].

While tin complexes of citrate [29] and malate [64] have specifically been implicated in poisoning incidents, the symptoms of acute tin poisoning (nausea, abdominal pain, vomiting, etc.), are very similar to those caused by staphylococcal toxin [43]. Thus, with hindsight, many cases of food poisoning said to be due to excess tin, may have been, in fact, of bacterial origin.

The process by which bulk, trace and contaminating elements are absorbed from the diet *in vivo* is imperfectly understood but it is clear that, in the absence of an active pumping mechanism, they diffuse through the gut wall down a concentration gradient. The absorption of Sn(II) from the alimentary tract of rats is reported to be 2.85% [67] or 7.65% [69] of the dose (when given as tin salts to fasted animals) whereas only 0.64% of the adminstered Sn(IV) is absorbed [67]. The difference between stannous and stannic uptake is presumably due to the greater insolubility of polyvalent cation complexes [70] and experiments have demonstrated that the tin is largely absorbed and transported in the oxidation state in which it was ingested [67].

An important tenet of chelation therapy and drug adminstration [71] is that neutral complexes traverse the hydrophobic phospholipid membrane of the intestine far more readily than charged species. The modelling of the speciation of tin in canned foods and toothpastes suggests that only a small percentage of the total tin ingested would be in the form of low molecular weight, soluble, neutral species suitable for absorption through the mucous membranes. This supports the well established observation that tin salts have low oral toxicity because they are poorly absorbed. In Bowen's classification [70], tin is given as a class II element (one which is absorbed to the extent of 5-7% across the gut wall). More recently however, it has been reclassified as type III (less than 5% absorbed [67,72].) It is interesting to note that, in contrast to the accepted hypothesis that only uncharged species are bioavailable, De Groot et al [42] have suggested that toxic effects may be induced by certain cationic tin compounds. The transport of these charged species is presumably due to the ability of living cytoplasm, which contains large quantities of immobile anions (nucleic acids, proteins, etc) to function as an efficient cation exchanger [70]. However, even if the cationic complexes were bioavailable, the speciation modelling predicts that very little of the tin is bound in this form.

The half-life of tin in the body has been estimated at values from 4 months [73] to 3 days (in the monkey [72]) but the most acceptable figure is 20-40 days [67,71]. Tin shows none of the cumulative toxicity of its group analogue lead, and the leaching of lead from the solder of a can seam may represent agreater biohazard than tin intoxication (stannous chloride is sometimes added to canned food to prevent the release of metals such as lead [3]).

Tin has a wide but variable distribution in human tissues with significant differences related to geographical location and with the lungs, to age [4]. The accumulation in the lungs is said to be due to inhalation rather than ingestion [67], and extreme chronic inhalation is known to produce a rare benign and symptomless pneumoconiosis, stannosis [19]. Of the small quantity of tin retained in the body, most is in the skeleton with smaller amounts in the liver and kidneys and some is fixed in the cells of the small intestine [67]. Very little is found in the brain, presumably because it is unable to cross the bloodbrain barrier [4] (unless in the form of organotin) and, because of low transplacental transfer [4] it is not fetotoxic. The exact form in which the metal is stored in the tissues is unknown but it is said to be lipid-extractable [74]. The urinary concentration of tin was formerly used as an index of absorption (on the basis that absorption and excretion were in balance) but it has now been established that a significant quantity of stannous (but not stannic) ions are excreted in the bile [67].

4.3 Toxicity and Essentiality

While chronic or acute poisoning from the consumption of

canned food is extremely remote, and the present recommended limit (200 ppm [9]) offers some margin of safety, in exceptional circumstances, high levels of tin may present an indirect threat. Tin is said to have a destructive action on blood corpuscles [75] and, in common with other heavy metals, it may cause the degradation of heme [76]. It could interact antagonistically with the absorption and metabolism of essential metals, particularly iron [42], copper and zinc [3] (it should be noted that these particular metals have been used in this study as chemical analogues for tin because of the similar charge: ionic radius ratio and there is likely to be competitive inhibition of their absorption at high concentrations of ingested tin). In addition, it is said to depress the absorption of selenium in man and if the diet contains little selenium, this may be significant [77]. Tin is reported to have a deleterious effect on the central nervous system [20] and, as mentioned previously, could inhibit protein digestion. In experiments with rabbits, Stone et al [78] have shown that stannous fluoride and chloride interfere with the processes of inflammation causing pustules to develop in traumatized skin, but the salts have no effect on undamaged tissue.

In an isolated report, Franke [76] linked cancers of the digestive - urinary system with tin levels in air and suggested a possible connection with tumours of the breast, lung and liver. Although it is a well established paradox that antitumour agents (which some tin compounds certainly are) may initiate cancers in healthy cells and despite the fact that canned foods have been indescriminately accused of causing cancer [79], no causal relationship has been established between tin and tumours. Indeed, even the implantation of tin foil into the subcutaneous tissues of rats does not produce a tumour whereas implants of several other metals produce sarcomata [8].

Although the concentrations of many trace elements (notably zinc) are high in the area of a tumour, tin levels are significantly lower than in surrounding non-neoplastic tissue and it has been noted that tin shows the greatest variation in concentration between normal and cancerous tissue of all known trace metals in man [80]. Various isolated reports have suggested that steel workers involved in tin-finishing operations have a decreasd incidence of cancer while populations living on tin-poor soils have an increased risk [80]. Tin is said to have an affinity for the thymus (as iodine does for the thyroid) and it has been suggested that the thymus produces tin biochemicals which act as anticarcinogens and/or antioncogens. It may be that in mammals, tin materials function as cell growth retardants (tin citrate inhibits the growth of Cl. botulinum [81]) and zinc compounds act as growth promoters (the markedly different concentrations in and around a tumour being evidence for this) [80].

The essentiality of tin as a trace metal in man is not proven and no deficiency state is recognised. Apart from the paramount role of carbon in living matter, and the role of silicon as a structural building material in plants, there is little evidence that the other Group IVA elements provide any essential biological service [82]. Tin is said to be an essential micronutrient in the rat [5], to stimulate growth in plants [5] and lichens are reported to concentrate it [74]. Schwarz [5] has pointed out that tin has a number of chemical properties which offer possibilities for biological functions. It forms complexes with variable co-ordination number and could contribute to the tertiary structure of proteins. It readily forms covalent bonds with carbon and since the potential of the stannous - stannic half cell is well within the physiological range (-0.15V), it could participate in electron transport reactions.

It has been said that the polluting metal cadmium, released into the environment in large quantities by zinc mining, may now be regarded as a beneficial micronutrient [71], and the same may be true of the originally non-ubiquitous tin.

Despite the fact that there is no direct experimental method of determining the speciation of inorganic tin in foods or biofluids, a promising approach has been to couple the capability of chromatography for separation with the selectivity and sensitivity of atomic spectrometry for detection. Thus, in recent years, Ebdon et al [83] have been able to separate a mixture of stannous, stannic and tributyltin in natural waters using coupled HPLC - flame atomic absorption spectrometry and Brinckman et al [84] have separated a mixture of organotins using a coupled HPLC - graphite furnace atomic absorption technique.

Although the primary source of tin in the diet is canned food, it enters the foodchain via several different routes. Tin compounds have been used as stabilizers in the manufacture of plastic bottles and films [17] and thence may leach into foods (quantities of the metal may be detected in orange juice and vinegar held in PVC containers [85]). It may be a constituent of phosphate fertilizers [4]; bottle caps [53] or, as described, of toothpastes. Tin may migrate from tinfoil used for wrapping foods [4] and, cheese in particular can accumulate up to 2000 ppm tin from this source [20] (and also may become contaminated with antimony [53]). It would be useful to establish the speciation of tin from these diverse sources.

5. CONCLUSIONS

A   In direct contrast to the organotins, which are known to be extremely poisonous, inorganic tin is largely non-toxic and the tin in canned foods is in the form of inorganic complexes
B   Although tin is an extremely common metal, almost universally ingested, very few unequivocal cases of chronic tin poisoning can be cited. This must be related to the facts that it is poorly absorbed, rapidly excreted and does not exert any profound biological effects.
C   It is of importance that the public should perceive the significance of bioavailability and the dramatic difference which may exist between *intake and uptake* of a given element.

The proportion of tin bound in soluble low-molecular-weight complexes in canned foods appears to be small.

D   The presence of tin in canned foods is a consequence of the ability of tin in tinplate to protect iron from corrosion (at its own expense) and thereby prevent spoilage of the food. In this sense it may be thought of as a necessary component of canned products.

E   The study of aqueous solutions of tin has been hindered by experimental problems of hydrolysis, precipitation and oxidation. Any further understanding of the speciaiton of tin in food and other materials would benefit greatly from improved experimentally determined stannous formation constants.

ACKNOWLEDGEMENTS

Two of us (DMM and JAV) wish to thank the Science and Engineering Research Council for studentships. We gratefully acknowledge the assistance of Ministry of Agriculture, Fisheries and Food, U.K. and Unilever Research, Port Sunlight Laboratories, U.K.

REFERENCES

[1]   J.W.Mellor, A Comprehensive Treatise on Inorganic and Theoretical Chemistry, **7**, Longmans, (1957)
[2]   A.O'Reardon Overbeck, National Geographic, 659, Nov (1940)
[3]   S.G.Schäfer and U.Femfert, Regul.Toxicol.Pharmacol., **4**, 57, (1984)
[4]   H.A.Schroeder, J.J.Balassa and I.H.Tipton, J.Chron.Dis, **17**, 483, (1964)
[5]   K.Schwarz, D.B.Milne and E.Vinyard, Biochem.Biophys.Res.Comm., **40**, 1, 22, (1970)
[6]   E.Mesk, C.R.Acad.Sci. (Paris), **176**, 138 (1923)
[7]   J.C.Sherlock, Int.Tin Res.Inst.Publ., **660**, 439 (1984)
[8]   M.Walters and F.J.C.Roe, Fd.Cosmet.Toxicol., 3, 271 (1965)
[9]   Food Additives and Contaminants Committee Report on the Review of Metals in Canned Foods. FAC/REP/38. HMSO London 1983.
[10] N.M.Brown, Ph.D. Thesis, Clemson University, (1972)
[11] A.J.Crowe, Int.Tin Res. Inst. Publ. No.676
[12] A.Frouin, Pr. Med., 25, 402 (1917)
[13] A.Frouin and R.Gregoire, C.R.Acad.Sci. (Paris), **164**, 794 (1917)
[14] R.Gregoire and A.Frouin, Bull.Acad.Med. (Paris), **77**, 704 (1917)

[15] J.A.Kolmer, H.Brown and M.J.Harkins, J.Pharmacol., **43**, 515 (1931)
[16] J.T.Rico, C.R.Soc.Biol. (Paris), **90**, 1098 (1924)
[17] J.M.Barnes and H.B.Stoner, Pharmacol.Rev., **11**, 211 (1959)
[18] Br.Med.J., **1**, 515 (1958)
[19] G.Pressel Tin in Handbook on Toxicity of Inorganic Compounds, Eds. Seiler, Sigel and Sigel, Marcel Dekker (1988)
[20] F.Browning, Toxicity of Industrial Metals, 2nd Ed, London, Butterworths (1969)
[21] S.J.Blunden, P.A. Cusack and R.Hill, The Industrial Uses of Tin Chemicals, Royal Society of Chemistry, London (1985)
[22] P.Gjermo. K.L.Baastad and G.Rolla, J.Peridontol.Res., **5**, 102 (1970)
[23] K.G.Yost and P.J.Van Demark, Appl.Environ. Microbiol., **35**, 920 (1978)
[24] N.Tinanoff, J.M.Brady and A.Gross, Caries Res., **10**, 415 (1976)
[25] A.Attramadal and B.Svatun, Scand. J.Dent.Res., **92**, 161 (1984)
[26] Scherer Laboratories Inc., "Gel-Kam" Prevent.Dent.Rev., **4**, 2 (1981)
[27] W.R.Wolf in The Importance of Chemical "Speciation" in Environmental Processes, Eds., M.Bernhard, F.E.Brinckman and P.J.Sadler, Springer-Verlag (1986)
[28] J.C.Westall, J.L.Zachary and F.M.M.Morel, Tech. Note No.18, Dept.Civil Engineering, MIT, Cambridge, MA (1976)
[29] Y.Omori, 7th Pacific Science Congress Proc., Tokyo 8, 11 (1966)
[30] A.I.Vogel, Textbook of Quantitative Inorganic Analysis, Longman, 4th Ed, 389 (1981)
[31] M.Filella and D.R.Williams, Inorg.Chim.Acta., **106**, 49 (1985)
[32] P.M.May, K.Murray and D.R.Williams, Talanta, **32**, 483 (1985)
McCance and Widdowson's, The Composition of Foods, HMSO (1979)
[34] The Canned Food Reference Manual, American Can Co. (1949)
[35] A.E.Martell and R.M.Smith in Critical Stability Constants, Vol.3, Plenum Press, New York (1977)
[36] A.R.Willey, Br.Corros.J., **7**, 29 (1972)
[37] J.C.Sherlock and S.C.Britton, Br.Corros.J., **7**, 180 (1972)
[38] J.Duffield, I.Kron, C.Morris and D.Williams, In preparation
A.Attramadal and B.Svatun Acta.Odontol.Scand, **38**, 349 (1980)
[40] C.A.Saxton, G.J.Harrap and A.M.Lloyd, J.Clin.Periodontol., **13**, 4, 301 (1986)
[41] D.J.Krutchkoff, T.H.Jordan, S.H.Y.Wei and W.D.Nordquist, Archs.Oral Biol., **17**, 923 (1972)
[42] A.P.De Groot, V.J. Feron and H.P.Til, Fd.Cosmet.Toxicol., **11**, 19 (1973)

[43] H.Cheftel Joint FAO/WHO Food Standards Program, Fourth Meeting of the Codex Committee on Food Additives, The Hague September (1967)
[44] A.Bakal and H.C.Mannheim, Israel J.Technol., **4**, 4, 262 (1966)
[45] I.Saguy, H.C.Mannheim and H.Passy, J.Food Technol., 8, 2, 147 (1973)
[46] S.Nagy, R.Rouseff and S.Ting, J.Agric.Food Chem., **28**, 1166 (1980)
[47] R.Rouseff and S.Ting J.Food Sci., **50**, 333 (1985)
[48] J.F.Kefford, H.A.McKenzie and P.C.O. Thompson, J.Sci.Food Agri., **10**, 51 (1959)
[49] S.G.Capar and K.W.Boyer, J.Food Safety, **2**, 105 (1980)
[50] M.Mahadeviah, Indian Food Packer, **30**, 2, 9 (1976)
[51] B.C.Goss, J.Ind.Eng.Chem., **9**, 2, 144 (1917)
[52] B.C.Goss, J.Biol.Chem., **30**, 53 (1917)
[53] S.Back, Food Manufacture, 381, November (1933)
[54] W.D.Bigelow, J.Ind.Eng.Chem., **8**, 9, 813 (1916)
[55] M.Mahadeviah, R.V.Gowramma, W.E.Eipeson and L.V.L.Sastry, J.Sci.Fd.Agric., **26**, 821 (1975)
[56] D.W.Gruenwedel and R.K.Patnaik, Chem.Mikrobiol.Technol.Lebensm., **2**, 97 (1973)
[57] D.W.Gruenwedel and H.C.Hao, J.Agric.Food Chem., **21**, 246 (1973)
[58] P.A. Cusack, P.J.Smith and J.D.Donaldson, J.C.S.Dalton, 439 (1982)
[59] W.T.Hall and J.J.Zuckerman, Inorg. Chem., **16**, 1239 (1977)
[60] T.N.Sumarokova, D.E.Surpina and L.V.Levchenko, Izv.Akad.Nauk Kaz.SSR,Ser.Khim., **18**, 26 (1968)
[61] T.Horio, Y.Iwamoto and I.Shiga, Int.Bottler and Packer, 54, December (1967)
[62] P.Fritsch, G.De Saint Blanquat and R.Derache Fd.Cosmet.Toxicol., **15**, 147 (1977)
[63] C.J.Benoy, P.A.Hooper and R.Schneider, Fd.Cosmet.Toxicol., **9**, 645 (1971)
[64] A.P.Luff and G.H.Metcalfe, Br.Med.J., **1**, 833 (1890)
[65] S.Warburton, W.Udler, R.M.Ewert and W.S.Haynes, Public Health Reports, **77**, 9, 798 (1962)
[66] W.Salant and J.B.Rieger, Proc.Soc.Exp.Biol.Med., **11**, 178 (1913)
[67] R.A.Hiles, Toxicol.Appl.Pharmacol., **27**, 366 (1974)
[68] W.Salant, J.B.Rieger and E.L.P.Treuthardt, J.Biol.Chem., **17**, 265 (1914)
[69] S.Kojima, K.Saito and M.Kiyozumi, Yakugaku Zasshi, **98**, 495 (1978)
[70] H.J.M.Bowen, Trace Elements in Biochemistry, Academic Press (1966)
[71] D.R.Williams and B.W.Halstead, J.Toxicol.:Clin.Toxicol., **19**, 10, 1081 (1982)
[72] J.E.Furchner and G.A.Drake, Health Phys, **31**, 219 (1976)
[73] J.G.Hamilton, Univ.Calif.Rad.Lab.Med.Health Phys.Quart.Rep., UCRL - 270, 4 (1948)
[74] E.J.Underwood, Trace Elements in Human and Animal Nutrition, 4th Ed., Academic Press, New York (1977)

[75] S.B.Schryver, J.Hyg., **9**, 3, 17 (1909)
[76] R.H.Franke, Annual Meeting of the Americal Association for the Advancement of Science, Washington, Feb 14 (1978)
[77] J.L.Greger, S.A.Smith, M.A.Johnson and M.J.Baier, Biol.Trace Element Res., **4**, 269 (1982)
[78] O.J.Stone and C.J.Willis, Toxicol.Appl.Pharmacol., **13**, 332 (1968)
[79] House of Lords Official Report, London **232**, 93, 327 (1961)
N.F.Cardarelli, B.M.Cardarelli and M.Marioneaux J.Nutrition, Growth and Cancer, **1**, 183 (1983)
[81] W.J.Scott and D.F.Stewart, J.Counc.Sci.Ind.Res.Aust., **17**, 16 (1944)
[82] K.Schwarz, Fed.Proc., **33**, 6, 1748 (1974)
[83] L.Ebdon, S.J.Hill and P.Jones, Analyst, **110**, 515 (1985)
[84] F.E. Brinckman, W.R.Blair, K.L.Jewett and W.P.Iverson, J.Chromatog.Sci., **15**, 493 (1977)
[85] L.H.Adcock and W.G.Hope, Analyst, **95**, 868 (1970)

Cellular interactions of organotin compounds in relation to their antitumor activity

André H. Penninks, Marianne Bol-Schoenmakers and Willem Seinen.
Research Institute of Toxicology
University of Utrecht
P.O. Box 80.176
NL-3508 TD Utrecht
The Netherlands

Introduction

From numerous studies it is evident now that some organotin compounds exhibit antitumor activity as was tested in particular against the P388 lymphocytic leukemia in mice (see for review, Crowe 1987). Special attention has been paid at first to the dialkyl- and diaryl-substituted organotin compounds, with dihalide or dipseudohalide complexes as anionic residues (Crowe et al 1980; 1984). More recently these anioic residues were replaced by organic molecules attached to the tin by either Sn-O, Sn-N or Sn-S bonds (Barbieri et al 1982; Saxena and Tandon 1983; Haiduc et al 1983; Huber et al 1985; Ruisi et al 1985; Gielen et al 1986a; 1986b). The structure activity relationships of the various diorganotin compounds tested revealed that in vivo the diethyl- and diphenyl- derivatives were the most promising (Crowe 1987). In the murine P388 prescreening test the best T/C values obtained did not reach values higher than 200 (Crowe et al 1987), values which are still lower than those of 200-300 observed with platinum derivatives in the same tumor system (Rosenberg et al 1969; Rosenberg 1979). In solid tymor systems only limited data are available on the effects of organotins. Crowe (1984) reported that many of the organotin complexes which were active towards P388 were found to be inactive in some solid tumor systems tested. In contrast, Cardarelli (1984a, 1984b), Arakawa (in press) and Penninks (1986) reported some activity of organotins towards solid tumors, although different tumor systems and/or routes of administration were used.

In regard to the mode of action of antitumor active organotin compounds almost no data are available obtained with tumor cells. Based on studies of Crowe et al (1984) it seems unlikely that organotin compounds will interact with DNA by the formation of cross-links of Sn with suitable orientated nitrogeneous bases, as seems to be a fairly well established explanation for the antitumor activity of cisplatin and its analogs (Prestayko et al 1980; Sherman et al 1985). Since organotin compounds are found to be very active in inhibiting tumor cell growth in vitro (Penninks et al 1986a, 1989; Gielen et al 1986b), a direct effect on malignant cell growth is likely to be involved in their in vivo antitumor activity. Our current knowlegde on cellular effects of organotins is mainly restricted to studies with cells of the immune system, because of the predominant immunotoxic effects of various di- and trisubstituted organotin homologs (Seinen and Penninks 1979; Snoeij et al 1987). In this paper comparable studies of a series of di- and trisubstituted organotin compounds (mostly halides) on cellular effects of in particular thymocytes will be presented. Besides the direct cytotoxic potentials, effects on cell energetics, macromolecular synthesis and membrane associated functions will be overviewed. The relation of these in vitro obtained cellular effects will be discussed in respect to the in vivo antitumor activity of the alkyl- and aryl substituted organotin halides, pseudohalides and Sn-O, Sn-S and Sn-N complexes.

Direct cytotoxic effects

To assess the cytotoxic effects of xenobiotics an array of test models is used in measuring parameters such as cell count, membrane damage, cell detachment, proliferation and metabolic activity. In general the cytotoxic activity of xenobiotics will differ depending the cell system-, culture medium- and concentration of the compounds[*] used as well as the time of incubation. Therefore it is not always easy to compare the activities of compounds obtained from different studies. Using the uptake of dyes as parameter for membrane damage the structure- activity relationships for di- and trialkyltin chlorides have been studied in various cell systems. In short term experiments of up to 4 hr cytotoxicity towards isolated rat thymocytes started above concentrations of 5 µM DBTC and 10 µM DETC. The most hydrophilic and lipophilic compounds tested, DMTC and DOTC respectively, were much less effective and showed cytotoxicity above levels of 120 µM only (Penninks and Seinen 1980). An equal cytotoxic structure-activity relation of the same dialkyltins was observed if thymocytes were exposed for up

---

[*] Abbreviations: DMT(C), dimethyltin(dichloride); DET(C), diethyltin(dichloride); DPrT(C), di-n-propyltin(dichloride);DBT(C), di-n-butyltin(dichloride); DPT(C), di-n-pentyltin(dichloride); DHT(C), di-n-hexyltin(dichloride); DOT(C), di-n-octyltin (dichloride); DPhT(C), diphenyltin (dichloride); TMT(C), trimethyltin(chloride); TET(C), triethyltin(chloride); TPrT(C), tri-n-hexyltin(chloride); TOT(C), tri-n-octyltin (chloride), TPhT(C), triphenyltin(chloride).

to 24 hr (Seinen et al 1977a; Arakawa and Wada, in press). Also for other mammalian cell lines i.e. Balb/c mouse 3T3 fibroblasts, mouse neuroblastoma N_2a cells and the fish BF_2 fibroblast cell line, the same sequence of cytotoxicity for diorganotin compounds was established (Borenfreund and Babich 1987; Babich and Borenfreund 1988). In the BF_2 cell line, isolated from the Bluegill sunfish, the cytotoxic potential of DPhT- and dicyclohexyltin chloride (EC_{50} value of 0.3) were found to be somewhat less and more effective than DBTC respectively (Table 1). In particular the studies with the BF_2 cell line demonstrate a higher activity with increasing chainlength indicating that lipophilicity appears to be the predominant factor in the membrane damaging properties of the dialkyltin compounds, as described by their logarithmic octanol/water partition coefficients (logP) and the Hansch π parameter of the organic ligand to the tin (Table 1).

Table 1. The EC_{50} cytotoxicity values for a series of di-substituted organotins in various cell systems in relation to their lipophilicity (logP) and hydrophobicity (Hansch π)*.

Compound	logP [a]	Hansch π parameter [b]	Cell viability (EC_{50}**, µM)	
			rat thymocytes [c]	BF2-fish fibroblasts [d]
DMTC	-3.10	0.56	> 120	40
DETC	-1.40	1.02	> 120	17
DPrTC	-	1.55	n.d.	11
DPhTC	1.90	1.96	n.d.	1
DBTC	1.49	2.13	6-10	0.8
DOTC	-	2.51	≥ 100	n.d.

* Cell viability was scored after 24 hr by means of a trypan exclusion test (c) or a neutral red assay (d)
** EC_{50}-effect concentration causing 50% cytotoxicity
    a - logarithmic n-octanol/water partition coëfficient according to Wong et al (1982).
    b - hydrophobicity of the organic ligand attached to the tin according to Laughlin et al (1985).
    c - adapted from Seinen et al (1977a), Vos et al (1984b).
    d - adapted from Babich and Borenfreund (1988).

Cytotoxicity of trisubstituted homologs was also studied by various authors (Snoeij et al 1986a; Arakawa and Wada, in press). The results of the studies of Arakawa, which were performed at a rather high dose level of 10^{-4} M only, showed and almost equal reduction of thymocyte viability after 24 hr for TETC, TBTC and TPhTC, whereas TMTC was somewhat less active. Snoeij et al (1986a) studied the cytotoxic capacity of a series of trisubstituted compounds in a dose (up to 8 µM) and time dependant way (up to 30 hr). The most active homologs appaered to be TBTC and TPhTC reducing dye exclusion of thymocytes within 2.5 hr at a concentration of 8 µM. TMTC and TETC did not affect this parameter until 10 h of incubation. In table 2 the structure-activity relationship is presented of the trisubstituted organotins inducing a 50 % reduction of thymocyte viability after 24 hr. The sequence of potency of this series was TETC>TBTC, TPhT>TMTC.

Table 2. The EC_{50} cytotoxicity values for a series of trisubstituted organotins for rat thymocytes in relation to their lipophilicity*

Compound	logP[a]	Thymocyte viability EC_{50} value (µM)
TMTC	-2.3	6
TETC	-1.8	2
TBTC	2.6	4
TPhTC	2.65	4

* thymocyte viability was scored after 24 hr by a trypan blue exclusion test (adapted from Snoeij et al 1986a)
[a] logarithmic n-octanol/water partition coefficient according to Wong et al (1982)

The cytotoxic activity of trisubstituted organotin compounds towards other cell species was found to differ in some extend as was also observed with the dialkyltins. Equal experimental conditions showed that rat bone marrow cells were somewhat less sensitive than rat thymocytes, whereas lysis of red blood cells of rats appeared at relative high doses (Snoeij et al 1986a), as was also described by others with red blood cells of many animal species (Byington et al 1974)

From the cytotoxicity data, measured by the uptake of dyes for membrane damage, it can be concluded that the structure-activity relations (SAR) of di- and trisubstituted organotins indicate a positive correlation between cytotoxicity and lypophilicity. Although the sensitivity of various cell systems differ in some extend the SAR within the series of di- and triorganotins was found to be the same. As was especially studied with butylated organotins (MBTC, DBTC, TBTC and TeBT)* there is a lack of correlation between either the degree of butylation or lipophilicity of the molecules and their in vitro cytotoxicity (Arakawa, in press; Babich and Borenfreund, 1988). This finding will be indicative for differential mechanisms of cytotoxicity.

---

* Abbreviations: MBTC, mono-n-butyltin trichloride; TeBT, tetra-n-butyltin.

Disturbances of mitochondrial energetics in vitro

Already in the first investigation on the biochemical effects of organotin compounds, diethyl- and triethyltin sulphate were recognized as powerful metabolic inhibitors, each with a different mode of action (Aldridge and Cremer 1955). At that time the toxicity of the organotin compounds was roused particularly by a tragic incident with a preparation, Stalinon, that was sold in France for the treatment of Staphylococcal skin infections. Diethyltin diiode was considered to be the active component but most likely due to contamination with triethyltin iodide, 217 people were poisoned in 1954, of which at least 100 of them died (Alajouanine et al 1958; Barnes and Stoner 1959).

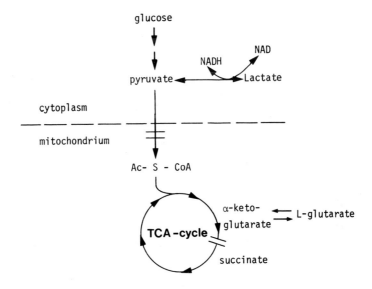

Fig.1. Schematic presentation of the inhibition of TCA cycle activity by dialkyltins, due to inhibitions of the mitochondrial pyruvate- and α-ketoglutarate dehydrogenase systems.

Diethyltin and other dialkyltin homologs (from dimethyl to dihexyltin) were found to inhibit oxygen and substrate consumption of isolated rat mitochondria (Aldridge 1976; Penninks and Seinen 1983). In the presence of dialkyltin compounds various substrates were not completely oxidized by the mitochondrial tricarboxylic acid cycle (TCA-cycle), but an accumulation of the α-keto acids pyruvic acid and α-keto glutarate was observed. The dialkyltin compounds were proposed to inhibit the two α-keto acid oxidizing enzyme complexes in mitochondria, namely pyruvate and α-keto glutarate dehydrogenase (Fig. 1). In this respect dialkyltins behave in a similar way to phenylarsenious acid, which might be in agreement with the observations that both have high affinity for dithiols (Aldridge and Cremer 1955). Moreover, the dithiol compound 2,3-dimercaptopropanol was found to be able to reverse the progressive accumulation of α-keto acids. Because of the high chemical affinity of dialkyltin compounds for dithiols, the inhibition of the a-keto acid dehydrogenase systems is suggested to be due to binding to the coenzyme lipoic acid or the enzyme lipoyl dehydrogenase (Fig. 2). Both molecules contain dithiol groups which are essential factors in the oxidation of the α-keto acids. The sequence of effectiveness for the series of dialkyltins was DBT, DPrT>DET>DPT, DHT>DMT.

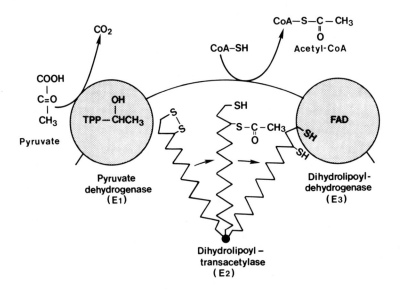

Fig 2. Schematic presentation of the α-keto acid dehydrogenase complex showing the conversion of pyruvate to acetyl-CoA, which will be further oxidized by the TCA cycle. The reduced dithiol group of lipoic acid (cofactor of E_2) or a dithiol group of the enzyme dihydrolipoyl dehydrogenase (E_3) are considered to be targets of dialkyltin interference.

At higher concentrations dibutyltin dichloride was also observed to inhibit the oxidative phosphorylation processes in mitochondria. Both an interaction with the ATP* synthase complex (Cain et al 1977) and an uncoupling effect have been described (Penninks et al 1983).

Extensive studies have also been performed in respect to the effects of triorganotins on mitochondrial function (Aldridge and Cremer 1955; Aldridge 1958, 1976; Aldridge and Street 1964, Selwyn et al 1970; Selwyn 1976). From these studies it was apparent that triorganotin compounds derange mitochondrial respiration in three different ways (Fig. 3), properties which were not shared by diorganotin compounds; (1) Triorganotins compounds mediated an exchange of halide for hydroxyl ions across mitochondrial membranes in halide containing media, resulting in a disturbance of the existing proton gradient which is of utmost importance for ATP synthesis; (2) Triorganotin compounds were also found to bind to a component of the ATP synthase complex, resulting in a direct inhibition of the mitochondrial ATP production; (3) Finally, particularly the more lipophilic triorganotin compounds interact with mitochondrial membranes leading to gross mitochondrial swelling and disruption.

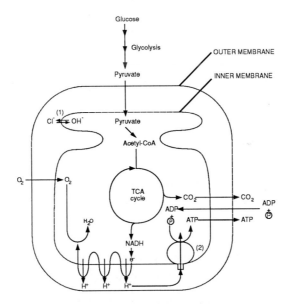

Fig. 3. Schematic presentation of the most important aspects of mitochondrial respiration. At (1) the trialkyltin induced exchange of Cl$^-$ versus OH$^-$ is indicated, which will affect the proton gradient built up by the electrontransport system, that is of utmost importance for ATP synthesis. At (2) the ATP-synthase complex is indicated. Triorganotins are considered to bind to a component of this complex, which will result in direct inhibition of ATP synthesis.

---

* Abbreviations: ATP, adenosine triphosphate.

As a result of these effects, the triorganotins act as effective inhibitors of mitochondrial ATP synthesis. The order of effectiveness of a series of trialkyltin compounds in causing a 50 % reduction of ATP production in isolated rat liver mitochondria was found to be TET>TPrT>TBT>THT>TMT (Aldridge 1976).

Disturbances of cellular energetics in vitro

To evaluate the significance of the organotin effects on mitochondrial respiration at the cellular level, comparative studies of a series of dialkyl- and trialkyltin chlorides were carried out using isolated rat thymocytes (Penninks and Seinen 1980, 1987; Snoeij et al 1986 a,b,c). Upon exposure to DMTC, DETC, DBTC and DOTC thymocyte glucose metabolism was studied, since from the mitochondrial studies it was apparent that α-ketoacid dehydrogenase complexes were inhibited (Fig. 1). A marked increase in the consumption of glucose, an accumulation of pyruvate and lactate and a reduced oxygen consumption were observed. Dibutyltin dichloride was the most active homolog in this respect, inducing a maximal stimulation of glucose consumption at a level of 5 µM (Penninks and Seinen 1980). Maximal stimulation of the other dialkyltin homologs are presented in table 3. Further increase of dialkyltin concentrations resulted in a rapid decline of glucose metabolism associated with a decrease in thymocyte viability. From the accumulation of lactate and pyruvate it is obvious that glucose is hardly metabolized oxidatively in the mitochondria. This was confirmed by the disturbed metabolism when lactate and pyruvate were used as oxidizable substrates (Penninks and Seinen 1980). In glucose containing media, cellular ATP levels were not found to be affected up to 5 µM DBTC, but decreased considerably when glucose was omitted from the medium (Penninks and Seinen 1987). These findings can be explained by considering the inhibition of pyruvate dehydrogenase caused by dialkyltin compounds. Due to this action the entrance of the glycolytic end product pyruvate into the TCA cycle is disturbed (Fig. 1). As a consequence, pyruvate accumulates in the cell and is largely converted into lactate in order to oxidize NADH to NAD*, which is one of the necessary cofactors of the glycolytic pathway. The resulting increased glucose consumption is considered to be an adaptation to the reduced TCA cycle activity which will lead to an activation of the glycolytic pathway. The increased glycolytic ATP

---

* Abbreviations: NAD, nicotinamide adenine dinucleotide; NADH reduced nicotinamide adenine dinucleotide.

phosphorylation is apparently capable, together with mitochondrial ATP formation from the oxidation of NADH and FADH*by the electron transport chain, to maintain the intracellular ATP levels, provided that glucose is present in the incubation medium.

Table 3. Maximal stimulation of glucose consumption of rat thymocytes incubated for 4 hr in the presence of a series of di- and trisubstituted organotin compounds.

Compound	µM	Compound	µM
DMTC	120	TMTC	slight effect up to 40
DETC	10	TETC	10
DPrTC	n.d.	TPrTC	2
DBTC	5	TBTC	1
DHTC	n.d.	THTC	1
DOTC	120	TOTC	no effect up to 10

n.d. not done

Thymocyte energetics was markedly inhibited upon incubation with various trialkyltins, except for the most hydrophilic and lipophilic homologs, trimethyltin and trioctyltin respectively (Snoeij et al 1986c). Incubation of rat thymocytes with concentrations higher than 10^{-7} M of TPrTC, TBTC or THTC resulted in an increased glucose consumption and a marked accumulation of lactate. TBTC and THTC were the most active inducing a maximal glycolytic activity at 1 µM. In contrast to the pyruvate accumulation induced by disubstituted organotins, the pyruvate levels were only slightly raised upon incubation with the active trisubstituted tin compounds. At higher dose levels stimulation of glucose consumption started to decrease again, probably due to membrane disrupting effects. Despite the stimulation of the glycolysis, incubation of thymocytes in phosphate buffered saline/glucose with TETC, TPrTC, TBTC or THTC resulted in a pronounced reduction in intracellular ATP levels, especially when glucose was omitted from the medium (Snoeij et al 1986b,c). In the presence of glucose ATP levels of thymocytes were already decreased to 66 % of the control value after an incubation period of 1 hr in the presence of 1 µM TBTC. Without glucose only 22 % of control ATP was left under the same conditions (Snoeij et al 1986b). The order of effectiveness to inhibit ATP synthesis in thymocytes differs considerable from the order of trialkyltin compounds disturbing ATP production in isolated mitochondria (see before). Not TETC is most effective in thymocytes

---

* Abbreviations: FADH, reduced flavin adenine dinucleotide.

but the more lipophilic homologs TPrTC, TBTC or THTC. Apparently to penetrate the cell and to interfere with mitochondrial respiration, the tin compound should not be too hydrophilic or too lipophilic.

The trialkyltin induced inhibition of ATP formation can be explained by the well-studied mode of action at the mitochondrial level. Both binding to the ATP synthase complex and mediating an ion exchange across the mitochondrial membranes will result in a reduced capacity to produce ATP. This latter phenomenon is responsible for a decreased pyruvate transport across the mitochondrial membranes, and a reduced activity of the oxidative phosphorylation, resulting in an cytoplasmic accumulation of pyruvate and NADH. At the expense of cytosolic NADH, pyruvate will be converted into lactate. The NAD produced will stimulate the glycolysis to continue with only a limitted production of ATP. In contrast to the dialkyltin compounds, the increased glycolysis does not provide enough ATP to meet the demands of trialkyltin-exposed cells. This is probably due to the different mode of action of trialkyltins which interfere with the activity of the TCA cycle and the electron transport chain. This will, in contrast to dialkyltins, not allow that NADH and FADH produced in the cell will be oxidized in the mitochondria resulting in a additional non glycolytic ATP production.

Disturbances of macromolecular synthesis in vitro

It was first shown in studies of Seinen et al (1979) that the blast transformation of rat thymocytes, in response to the mitogens PHA and Con A was inhibited in a dose-related manner in the presence of graded amounts of DBTC and DOTC. The ^{3}H-TdR incorporation into DNA, measured after 72 hr of incubation, was already completely inhibited at concentrations of 0.33 µM DBTC and 1.2 µM DOTC. Since freshly isolated thymocytes incorporate precursors for DNA, RNA and proteins at a considerable rate, without mitogenic stimulation, the effects of organotins on these parameters were extensively studied in short time experiments. Micromolar concentrations of DETC and DBTC effectively decreased the incorporation of DNA* and protein precursors, while the incorporation of ^{3}H-Urd

---

* Abbreviations: PHA, phytohemagglutinin; Con A, concanavalin A; ^{3}H-TdR, tritiated thymidine; DNA, deoxyribonucleic acid; RNA, ribonucleic acid; ^{3}H-Urd, tritiated uridine; ^{14}C-Leu, ^{14}C-labeled leucine.

into RNA increased by these dialkyltins (Miller et al 1980, Penninks and Seinen, 1983a;1983b;1987). The half-maximal inhibition concentrations (IC_{50}) for ^{3}H-TdR and ^{14}C-Leu incorporations of thymocytes in 1 hr incubations in the presence of DBTC (Table 4), indicate that protein synthesis is most sensitive. The increased ^{3}H-Urd incorporation upon incubation with DBTC seems to result from a remarkable stimulation found in the subfraction of non dividing small thymocytes (Penninks et al 1986b). The order of effectiveness of a series of dialkyltins was only studied in respect to the inhibition of DNA synthesis of rat thymocytes after a 72 hr incubation period (Penninks et al 1986a). DBTC was found to be the most effective sequenced by DOTC>DPhTC>DET>DMTC.

Table 4. The half-maximal inhibition concentrations of di- and tributyltin chlorides on DNA, RNA and protein synthesis of freshly isolated rat thymocytes*

Compound	EC_{50} values (µM) on the incorporation of		
	^{3}H-TdR	^{3}H-Urd	^{14}C Leu
DBTC[a]	3.5	4.6**	0.6
TBTC[b]	0.32	0.95	0.36

* After a 30 min preincubation with various concentrations of DBTC and TBTC, the radiolabeled precursors were added and the incorporation was measured after an additional incubation period of 60 min. ([a]Penninks et al 1987; [b]Snoeij et al 1986).

** Upon incubation with 4.6 µM DBTC, ^{3}H-Urd incorporation was stimulated 2 fold.

Upon incubation with TBTC the DNA, RNA and protein synthesis of thymocytes was found to be affected too (Snoeij et al 1986abc). As shown by the EC_{50} values presented in table 4, TBTC is far more effective in inhibiting the macromolecular synthesis of thymocytes than DBTC. Moreover in contrast to the stimulating effects of DBTC, TBTC inhibitid the ^{3}H-Urd incorporation of thymocytes. The structure activity to inhibit precursor incorporation of thymocytes decreased in the following order TBTC, THTC>TPTC>TETC>TMTC> TOTC (Snoeij et al 1986). Up to levels of 4 µM the most hydrophilic (TMTC) and lipophilic (TOTC) compounds were found to be inactive, whereas at levels of 0,5 µM of TBTC and THTC the precursor incorporation for DNA, RNA and protein synthesis were already severely diminished. The effects of TPhTC were only studied on ^{3}H-TdR incorporation of thymocytes after a 5 hr incubation period and found to be as active as TBTC and TPrTC (Snoeij et al 1986a). Antiproliferative effects of trialkyltin compounds were also described for other cell species e.g. a baby hamster kidney cell line (Reinhardt et al 1982), rabbit chondrocytes (Webber et al 1985) and rat skin, either in organ culture or in vivo (Kao et al 1983; Middleton and Pratt 1978).

Summary and conclusions on the cellular effects of organotins

Since freshly isolated thymocytes display a high basal proliferative activity, which also demands a continuous energy production, the effects of organotins on thymocytes might be of importance for their antitumor activity. Various aspects of the DBTC- and TBTC-induced disturbances in thymocyte metabolism, all performed under the same experimental conditions are summarized in Fig. 4.

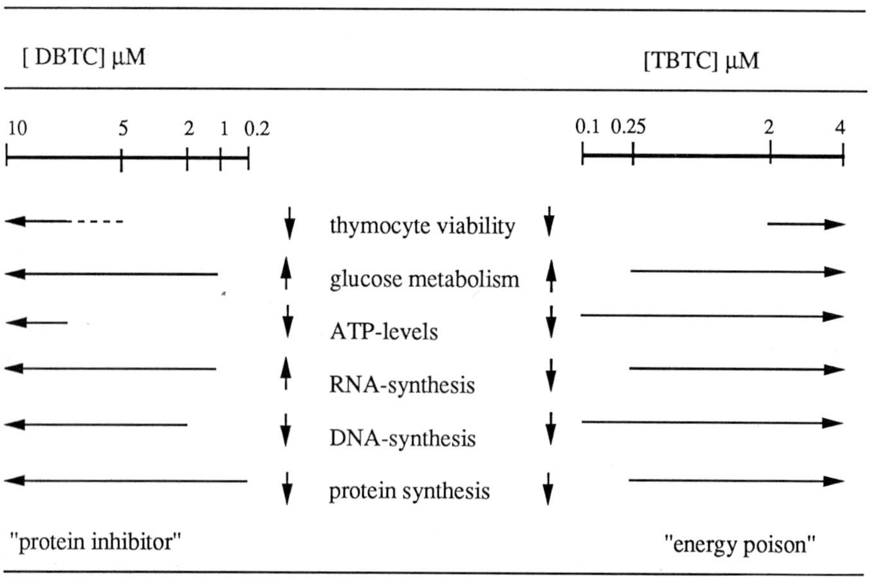

[DBTC] µM		[TBTC] µM
10 5 2 1 0.2		0.1 0.25 2 4
← ----	↓ thymocyte viability ↓	——→
← ——————	↑ glucose metabolism ↑	————→
← ——	↓ ATP-levels ↓	————→
← ————	↑ RNA-synthesis ↓	————→
← ————	↓ DNA-synthesis ↓	————→
← ——————	↓ protein synthesis ↓	————→
"protein inhibitor"		"energy poison"

Fig. 4 Summary of effects of DBTC and TBTC on cellular metabolism of rat thymocytes.*
*Results are based on 1 hr incubation studies with freshly isolated thymocytes, preceded with a 30 min pre-incubation period.

If the effective dose levels of DBTC and TBTC are compared it is apparent that TBTC is far more cytotoxic than DBTC. The direct cytotoxic activities measured by the dye exclusion test, are both for DBTC and TBTC observed at rather high dose levels. They will probably result from a combination of effects on the cellular metabolism and the direct membrane disrupting activities. Disturbances of cellular metabolism are in particular reflected by marked effects on glucose metabolism, cell energetics and macromolecular synthesis.

Since up to 5 µM DBTC thymocyte viability was not affected, the functional inhibition of thymidine and leucine incorporation and the stimulation of uridine incorporation can not be accounted for by cell loss or reduced viability. Futhermore, it seems unlikely that these effects are secundary to a limited energy supply, since up to 5 µM DBTC the ATP levels were not reduced provided that glucose was present. The mode of action of DBTC on thymocyte energetics resembles that observed in isolated mitochondria, and involves the inhibition of α-keto acid dehydrogenase enzyme complexes. As an adaptation to the diminished TCA cycle activity, thymocyte glycolysis will be activated as was reflected by the increased glucose metabolism. From the DBTC effects on macromolecular synthesis the stimulated ^{3}H-Urd incorporation was a somewhat unexpected finding. Additional studies have shown that the ^{3}H-Urd incorporation of a non proliferating subpopulation of thymocytes is responsible for this effect (Penninks et al 1986b). ^{3}H-Urd incorporation of the proliferating subpopulation of thymocytes was not significantly affected. Inhibition of protein synthesis was found to be most sensitive parameter. It was already reduced by 50% at a dose level of DBTC at which DNA synthesis was not diminished yet. Comparative studies with well known inhibitors of protein synthesis revealed that DBTC was an even more effective inhibitor of protein synthesis of thymocytes than cycloheximide and puromycin (Penninks, unpublished results). Moreover, all protein synthese inhibitors diminished the ^{3}H-TdR incorporation like was observed with DBTC. These results indicate that at low dose levels DBTC will be a very effective inhibitors of protein synthesis and that effects on ^{3}H-TdR incorporation might be secundary to the inhibition of protein synthesis.

For TBTC, the well known effects studied in isolated mitochondria are likely to be responsible for their severe effects on thymocyte energetics. The increased glucose metabolism is considered to be an adaptation to the diminished oxidative ATP generation. However, it does not meet the ATP needs of the cell. Membrane passage of precursors for macromolecular synthesis as well as phosphorylation of nucleosides i.e. thymidine kinase were not affected by TBTC (Snoeij et al 1986b,c). Although effects on the polymerisation of DNA, RNA and proteins have not been studied yet, it is postulated bij Snoeij et al (1986c) that TBTC is best characterized as an energy poison. The additional effects on macromolecular synthesis are considered to result from the disregulation of cellular energetics. Further support for this hypothesis was obtained by comparative studies with well known energy poisons. Besides marked effects on ATP levels, they also affected the macromolecular synthesis of thymocytes like was observed with TBTC (Snoeij et al 1986c).

In respect to the order of in vitro cytotoxicity for the series of di- and trisubstituted organotins the more hydrophilic (DMTC, TMTC) and the more lipophilic (DOTC, TOTC) homologs were in general less active. Homologs with intermediate chain length, propyl, butyl, hexyl as well as phenyl, were mostly more active than the more hydrophilic ethyl analogs. However, on some occasions the order of effectiveness was found to be different. This indicates that not only chemical properties of the respective organotin compounds (lipophilicity or hydrophilicity) but also the biophysical environment of the process studied, will determine which homolog is the most active.

Vivo-Vitro relation of antitumor activity of organotins

Although numerous studies have been performed regarding the antitumor potency of various organotin compounds only limited information is available concerning the mechanism(s) of their in vivo activity towards tumor cells. Most studies with organotins were directed to study their activity in the P388 prescreening tumor model, in order to obtain a T/C value that will give an impression of the activity of the tested compounds. However, this model has only limited value as to predict the activity of the tested compound towards solid tumors, which will be the most important objective for this kind of research. In vitro antitumor assays may in general present the same information as the P388 model for compounds with an expected direct activity towards tumor cells. The most important diorganotin structures studied for their in vivo antitumor activity are presented in table 5. The structure-activity relationships within each class of the compounds revealed that the ethyl- and phenylsubstituted homologs were found to be the most active (Crowe 1987). Dibutyltin homologs showed only limited activity, whereas dioctyltin derivates were mostly inactive. Only dioctyltin dichloride (Penninks et al 1986a) and dioctyltin glycylglycinate (Barbieri et al 1982) were reported to show some activity. With trisubstituted organotin complexes only limited studies have been performed in vivo. Although Brown (1972) and Cardarelli (1984) found some activity with TPhT and TBT compounds by oral application in the food or drinking water, no sufficient activity was observed in the P388 model to warrant further testing (Crowe 1980; Huber et al 1985)

Table 5. Most important organotin structures studied for their in vivo antitumor activity

	most active
- diorganotin halides-R_2SnX_2	R=ethyl or phenyl

- diorganotin dipseudohalides $R_2SnX_2.L_2$
   L_2- two monodentate ligands
   L_2- one bidentate ligand
	R=ethyl or phenyl

- diorganotin compounds containing Sn-O, Sn-N and Sn-S bonds	R=ethyl or phenyl

Comparative in vitro antitumor studies with organotin compounds are limited too. The effects of a series of di- and trisubstituted organotin chlorides on the DNA replication of various human leukemic cells were studied by Penninks et al (1986a). The order of effectiveness for both the di- and trisubstituted organotins was butylated and phenylated>ethylated>octylated>methylated. Moreover, in most cases the trisubstituted homologs were active at somewhat lower dose levels. Comparative in vitro screening of the DMT-, DBT- and DPhT- derivates of 2,6-pyridine dicarboxylic acid towards various mice leukemia cells revealed an almost equal activity for the butyl- and phenyl compounds whereas the methyl homologs were an order of 100-200 less active. Whether the activity of these stannylene derivates of 2,6-pyridine dicarboxylic acids were more active than their respective parent compounds could not be answered since these were not included in the study. Although limited in number these in vitro antitumor studies indicate that like observed in vitro with thymocytes, the butyl and phenyl homologs were the most active. They were even somewhat more active than diethyltins, which together with the diphenyl tin compounds were the most active compounds in vivo. So their is a discrepancy between the most active diorganotin homologs in in vivo antitumor assays (DET and DPhT) and the most active homologs in vitro (DBT, DPhT>DET) towards tumor - as well as normal cells. In the studies of Crowe et al (1980, 1984) it was indicated that the organotin compounds tested were administered as a solution in saline, saline and Tween 80 or as a suspension. Since it was not indicated in which vehicle the various structural homologs were tested, the existing differences in water solubility of the homologs might have played an important factor in finding DBT among the minor active homologs. Solutions of the relative best water soluble diethyltins might have been obtained, which can be doubted for the dibutyltins in particular at the extreme high dose levels up to 400 mg/kg (Crowe et al 1980a). Greatest caution should be exercised in comparing results obtained with a solution or suspension of structural analogs. The bio-availability and subsequent biological activity of the tin compounds will not only be determined by the administration in a solution or a suspension, but also by the type and amount of the solubilizer used. For further testing it will be very important to pay more attention in preparing solutions and/or suspensions of the organotins to be tested. Moreover it will be useful for comparative antitumor studies in vivo and in vitro to administer the various organotin derivates at equimolar dose levels. Another explanation for the observed low activity of the dibutyltins in vivo was presented by Sherman and Huber (1988). They suggested that without an active immune system the organotins would be less capable in preventing tumor cell growth. Since dibutyltin compounds were described to have a high thymolytic activity compared to the diethyl- and diphenyl homologs (Seinen et al 1977a; Seinen and Penninks 1979), a suppressed immune system was considered in case of the dibutyltins which could account for their lower antitumor activity. However, although the organotin induced thymus atrophy is a fast phenomena in rats (Snoeij et al 1988), disturbed immune functions could only be observed after prolonged exposure (Seinen et al 1977b,1979; Seinen 1981; Vos et al 1984a). Therefore it seems not likely that in the i.p. P388 leukemia model of 10-15 days, a possible immune suppression can explain the lower activity of dibutyltins versus diethyl and diphenyltins.

Mode of action of organotins in relation to their antitumor activity

Most antitumor active drugs exhibit a direct interaction with tumor cells by acting as antimetabolites at the cytosolic or nuclear level. Up to now there is no clear explanation for the antitumor activity of organotins. Based on their antiproliferative activity and cytotoxicity at low dose levels, as studied in tumor cells and thymocytes, a direct interaction seems to be involved. However, Cardarelli (1984 a,b) suggested that an indirect mechanism will account for the observed antitumor activity of organotins. In considering Sn to be a vital element for life (Swarz et al 1971; Cardarelli et al 1983) he suggested that a.o. Sn compounds administered in whatever form or route, are converted to anticarcinogenic organotin compounds in the thymus. Probably in a steroid form they will subsequently kill tumor cells or prevent the proliferation of tumor cells. Future studies will have to prove whether the observed antitumor activity of various butylated compounds upon long term oral application at low dose levels in the drinking water (Cardarelli et al 1984 a,b) will result from the presence in the circulation of the administered compound itself or from a tin-steroid produced in the thymus. To further study this hypothesis of Cardarelli, comparative antitumor experiments could be performed in normal and nude mice or rats, since the so called nude animals lack the presence of a thymus.

Concerning the mechanism(s) of a direct interaction of organotins with tymor cells only limited information is available. The organotin induced activation of thymocyte glycolysis could not be studied in tumor cells, due to a high anaerobic metabolism as was indicated by a high glycolytic activity (Penninks et al, unpublished results). In the same studies it was observed that after a 24 hr incubation the number of tumor cells in the presence of DBTC was less increased than in the control tumor cell suspensions, without an obvious loss of cell viability. This indicates that the reduced tumor cell number observed at low DBTC doses will not be due to a direct cytotoxic action, but will rather result from an antiproliferative activity, as was observed in various in vitro studies (Gielen et al 1986b; Penninks et al 1986a; 1989).

Since so many different organotin complexes were found to be active the question was raised what the active species of the organotins would be. For organotin halides it is supposed that they will easily dissociate in biological fluids. The resulting R_2Sn^{++} moiety is considered to be the active molecule, inwhich the lipophilicity of the R_2 together with the reactivity of the Sn^{++} will determine its activity. As is already discussed by Crowe (1987) there are a lot of indications to suppose that also the octahydral complexes with two mono or one bidentate ligand(s), as well as the Sn-O, Sn-N and Sn-*bonded complexes will dissociate to R_2Sn^{++} species. For the $R_2SnX_2.L_2$ complexes it was observed that their stability was determined by the length of the tin-ligand (L_2) bonds. The $R_2SnX_2.L_2$ activity increased with longer $Sn-L_2$ bonds, indicating that the predissociation of the bidentate ligand will be a crucial step for their activity. The activity of the Sn-O, Sn-N and Sn-S complexes seems also to depent on the hydrolysis of this bond. Hydrolytically stable compounds show less or no activity if compared with hydrolytically unstable compounds (Crowe 1987). Studies of Ruisi et al (1985) support this hypothesis, since they indicate that the activity of the dialkyltin glycylglycinates will probably result from the dialkyltin species. Moreover, the

---

[*] Abbreviations: R_2Sn^{++}, dialkyltin mioety inwhich R stands for various possible alkyl or aryl chains.

equal in vitro antitumor activity observed with DBTC and Sn-O bonded DBT-derivates of 2,6-pyridine dicarboxylic acid, pyridoxine, cortexolone and erythromycine, indicate the unstability of this bond even in vitro (Penninks et al 1989). Since the ligands bonded to the tin did not show additional toxicity if released, the stability before hydrolysis will rather determine the pharmacokinetic behaviour (transport to the side of action, transmembrane passage etc.) than the toxicity of the organotin complexes. At least for the Sn-O bonded DBT-derivatives of 2,6- pyridoxine dicarboxylic acid, pyridoxine, cortexolone and erythromycine it seems obvious that they are already released before entering the cell, since otherwise the different ligands would certainly have influenced cell uptake and the subsequent cytotoxicity

In conclusion our current knowlegde indicates that, in whatever intermediate form it will be presented to the cell, the R_2Sn^{++} is likely to be the ultimate reactive group. Although limited studies have been performed with trisubstituted compounds in this respect, its activity might also result from the R_3Sn^+ moiety. The actual mode of action towards tumorcells is still unclear, although an antiproliferative activity seems to be involved. A direct interaction of organotins with nitrogenous bases on DNA, like was observed for cisplatin and its analogs, was excluded (Crowe et al 1984). If cross-linking of DNA will still be involved than it will take place via a different route. Based on our current knowlegde, obtained with thymocytes, the antiproliferative activity of di- and trisubstituted compounds might also result from effects on protein synthesis and energy metabolism, respectively. Further studies will be needed to evaluate this hypothesis and to elaborate the interactions at the molecular level.

References

Alajouanine Th, Dérobert L, Thiélffry S (1958) Etude clinique d'ensemble de 210 cas d'intoxication par les sels organiques d'etain. Rev Neurol 98:85-96
Aldridge WN (1958) The biochemistry of organotin compounds. Trialkyltins and oxidative phosphorylation. Biochem J 69:367-376
Aldridge WN (1976) The influence of organotin compounds on mitochondrial functions. Adv Chem Ser 157:186-196
Aldridge WN, Cremer JE (1955) The biochemistry of organotin compounds. Diethyltin dichloride and triethyltin sulphate. Biochem J 61:406-418
Aldridge WN, Street BW (1964) Oxidative phosphorylation. Biochemical effects and properties of trialkyltins. Biochem J 91-287-297
Arakawa Y, Wada O. In: Zuckerman JJ (ed) Tin and malignant cell growth. CRC Press, Boca Raton, Florida, in press

Babich H, Borenfreund E (1988) Structure-activity relationships for diorganotins, chlorinated benzenes, and chlorinated anilines established with Bluegill Sinfish BF-2 cells. Fund and Appl Toxicol 10:295-301

Barbieri R, Pellerito L, Ruisi G, Lo Giudice MT (1982) The antitumour activity of diorganotin (IV) complexes with adenine and glycylglycine. Inorg Chim Acta 66:L39-L40

Barnes JM, Stoner HB (1959) The toxicology of tin compounds. Pharmacol Rev 11:211-232

Borenfreund E, Babich H (1987) In vitro cytotoxicity of heavy metals, acrylamide, and organotin salts to neural cells and fibroblasts. Cell Biol Toxicol 3:63-73

Brown NM (1972) The effect of two organotin compounds on C3H-strain mice, PhD thesis, Clemson University

Byington KH, Yeh RY, Forte LR (1974) The hemolytic activity of some trialkyltin and triphenyltin compounds. Toxicol Appl Pharmacol 27:230-240

Cain K, Hyams RL, Griffiths DE (1977) Studies on the energy-linked reactions: Inhibition of oxidative phosphorylation and energy-linked reactions by dibutyltin dichloride. FEBS lett 82:23-28

Cardarelli NF, Cardarelli BM, Libby EP, Dobbins E (1984b) Organotin implications in anticarcinogenesis. 2. Effects of several organotins on tumour growth rate in mice. Austral J Exp Biol Med 62:209-214

Cardarelli NF, Cardarelli BM, Marioneaus M (1983) Tin as a vital trace nutrient. J Nutr Growth Cancer 1:181-194

Cardarelli NF, Quitter BM, Allen A, Dobbins E, Libby EP, Hager P, Sherman LR (1984a) Organotin implications in anticarcinogenesis. 1. Background and thymus involvement. Austral J Exp Biol Med 62:199-208

Crowe AJ (1980) Synthesis and studies of some biologically active organotin compounds, PhD thesis, London University

Crowe AJ (1987) The chemotherapeutic properties of tin compounds. Drugs of the Future 12:255-275

Crowe AJ, Smith PJ, Atassi G (1980) Investigations into the antitumour activity of organotin compounds 1. Diorganotin dihalide and dipseudohalide complexes. Chem Biol Interact 32:171-178

Crowe AJ, Smith PJ, Atassi G (1984) Investigations into the antitumour activity of organotin compounds 2. Diorganotin dihalide and dipseudohalide complexes. Inorg Chim Acta 93:179-184

Gielen M, De Clercq L, Willem R, Joosen E (1986a) New developments in antitumour active organotin compounds In: Cardarelli (ed) Tin as a vital nutrient: implications in cancer prophylaxis and other physiological processes. CRC Press, Cleveland

Gielen M, Willem R, Mancilla T, Ramharter J, Joosen E (1986b) Strategy for the development of novel organotin anti-cancer agents. Si Ge Sn and Pb Comp 9:349-366

Haiduc I, Silverstra L, Gielen M (1983) Organotin compounds: New organometallic derivates exhibiting antitumor activity. Bull Soc Chim Belg 92:187-189

Huber F, Roge G, Carl L, Atassi G, Spreafico F, Filippeschi S, Barbieri R, Silvestri A, Rivarola E, Ruisi G, Di Bianca F, Alonzo G (1985) Studies on the antitumour activity of di- and tri-organotin (IV) complexes of amino acids and related compounds, of 2-mercaptoethane sulphonate, and of purine-6-thiol. J Chem Soc Dalton Trans 523-527

Kao J, Hall J, Holland JM (1983) Quantitation of cutaneous toxicity: An in vitro approach using skin organ culture. Toxicol Appl Pharmacol 68:206-217

Laughlin RB, Johannesen RB, French W, Guard H, Brinckman FE (1985) Structure-activity relationships for organotin compounds. Environ Toxicol Chem 4:343-351

Middelton MC, Pratt I (1978) Changes in incorporation of 3H-thymidine into DNA of rat skin following cutaneous application of dibutyltin, tributyltin and 1-chloro 2:4-dinitrobenzene and the relationship of these changes to a morphological assessment of the cellular damage. J Invest Dermatol 71:305-310

Miller RR, Harting R, Cornish HH (1980) Effects of diethyltin dichloride on amino acids and nucleoside transport in suspended rat thymocytes. Toxicol Appl Pharmacol 55:564-571

Penninks AH, Bol-Schoenmakers M, Gielen M, Seinen W (1989). A comparative study with di-n-butyltin dichloride and various Sn-O bonded dibutyltin derivates on the macromolecular synthesis of isolated thymocytes and the in vitro and in vivo antitumor activity. Main Group Metal Chemistry 12:1-15

Penninks AH, Punt PM, Bol-Schoenmakers M, van Rooijen HJM, Seinen W (1986a) Aspects of the immunotoxicity, antitumor activity and cytotoxicity of di- and trisubstituted organotin halides. Si Ge Sn and Pn Comp 9:367-380

Penninks AH, Seinen W (1980) Toxicity of organotin compounds. IV. Impairment of energy metabolism of rat thymocytes by various dialkyltin compounds. Toxicol Appl Pharmacol 56:221-231

Penninks AH, Seinen W (1983a) The lymphocyte as target of toxicity: A biochemical approach to dialkyltin induced immunosuppression. In: Hadden JW, Chedid L, Dukor P, Speafico F, Willoughby D (eds) Advances in immunopharmacology, Pergamon Press, Oxford

Penninks AH, Seinen W (1983b) Immunotoxicity of organotin compounds. In: Gibson GG, Hubbard R, Parke DV (eds) Immunotoxicity, Academic Press, London

Penninks AH, Seinen W (1987). Immunotoxicity of organotin compounds. A cell biological approach to dialkyltin-induced thymus atrophy. In: Berlin et al (eds). Immunotoxicology. Proceedings of the international symposium on the immunological system as target for toxic damage. Nijhoff, Hingham MA

Penninks AH, Snoeij NJ, Seinen W (1986b) Thymocytes as target of dialkyltin toxicity. In: Chedid L, Hadden JW, Spreafico F, Dukor P, Willoughby D (eds) Advances in Immunopharmacol, Pergamon Press, Oxford

Penninks AH, Verschuren PM, Seinen W (1983) Di-n-butyltin dichloride uncouples oxidative phosphorylation in rat liver mitochondria. Toxicol Appl Pharmacol 70:115-120

Prestayko AW, Crooke ST, Carter SK. (eds) (1980) Cisplatin: Current status and new developments, Academic, New York

Reinhardt CA, Schawalder H, Zbinden G (1982) Cell detachment and cloning efficiency as parameters for cytotoxicity. Toxicology 25:47-52

Rosenberg B (1973) Platinum coordination complexes in cancer chemotherapy. Naturwissenschaffen 60:399-406

Rosenberg B, Vancamp L, Trosko JE, Mansour VH (1969) Platinum compounds: A new class of potent antitumor agents. Nature 222:385-386

Ruisi G, Silvestri A, Lo Guidice MT, Barbieri R, Atassi G, Huber F, Grätz K, Lamartina L (1985) The antitumor activity of di-n-butyltin(IV)glycylglycinate, and the correlation with the structure of dialkyltin(IV)glycylglycinates in solution studies by conductivity measurements and by infrared, nuclear magnetic resonance and Mössbauer spectroscopic methods. J Inorg Biochem 25:229-245

Saxena A, Tandon JP (1983) Antitumour activity of some diorganotin and tin (IV) complexes Shiff bases. Cander Lett 19:73-76

Schwarz K (1971) Tin as an essential growth factor for rats. In: Meritz W, Cornatzer WE (eds) Newer trace elements in nutrition, Marcel Dekker, New York

Seinen W (1981) Immunotoxicity of alkyltin compounds. In: Sharma (ed) Immunological considerations in toxicology. CRC Press, Boca Roton

Seinen W, Penninks AH (1979) Immune suppresion as a consequence of a selective cytotoxic activity of certain organometalic compounds on thymus and thymus-dependent lymphocytes. Ann NY Acad Sci 320:499-517

Seinen W, Vos JG, Brands R, Hooykaas H (1979b) Lymphocytotoxicity and immunesuppression by organotin compouds. Suppression of GvH activity, blast transformation and E-rosette formation by di-n-butyltin dichloride and di-n-octyltin dichloride, Immunopharmacol 1:343-355

Seinen W, Vos JG, Van Spanje I, Snoek M, Brands R, Hooykaas H (1977a) Toxicity of organotin compounds. II. Comparative in vivo and in vitro studies with various organotin and organolead compounds in different animal species with special emphasis on lymphocyte cytotoxicity. Toxicol Appl Pharmacol 42:197-212

Seinen W, Vos JG, van Krieken R, Penninks AH, Brands R, Hooykaas H (1977b) Toxicity of organotin compounds. III. Suppression of thymus dependent immunity in rats by di-n-butyltindichloride and di-n-octyltin dichloride. Toxicol Appl Pharmacol 42:213-224

Selwyn MJ (1976) Triorganotin compounds as ionophores and inhibitors of ion translocating ATPases. Adv Chem Ser 157:204-226

Selwyn MJ, Dawson AP, Stockdale M, Gains N (1970) Chloride-hydroxide exchange across mitochrondrial, erythrocyte and artificial lipid membranes, mediated by trialkyl- and triphenyltin compounds. Eur J Biochem 14:120-126

Sherman LR, Huber F (1988) Relationship of cytotoxic groups in organotin molecules and the effectiveness of the compounds against leukemia. Appl Organomet Chem 2:65-72

Sherman SE, Gibson D, Wang AHJ, Lippard SJ (1985) X-ray structure of the major adduct of the antitumour drug cisplatin with DNA: cis-[Pt(NH3)2{d(pGpG)}]. Science 230:412-417

Snoeij NJ, Penninks AH, Seinen W (1987) Biological activity of organotin compounds - An overview. Environ Res 44:335-353

Snoeij NJ, Penninks AH, Seinen W (1988) Dibutyltin and tributyltin compounds induce thymus atrophy in rats due to a selective action on thymic lymphoblasts. Int J Immunopharmac 10:891-899

Snoeij NJ, Punt PM, Penninks AH, Seinen W (1986b) Effecs of tri-n-butyltin chloride on energy metabolism, macromolecular synthesis, precursor uptake and cyclic AMP production in isolated rat thymocytes. Biochim Biophys Acta 852:234-243

Snoeij NJ, Van Iersel AAJ, Penninks AH, Seinen W (1986a) Triorganotin-induced cytotoxicity to rat thymus, bone marrow and red blood cells as determined by several in vitro assays. Toxicology 39:71-83

Snoeij NJ, Van Rooijen HJM, Penninks AH, Seinen W (1986c) Effects of various inhibitors of oxidative phosphorylation on energy metabolism, macromolecular synthesis adn cyclic AMP production in isolated rat thymocytes. Biochim Biophys Acta 852:244-253

Vos JG, de Klerk A, Krajnc EI, Kruizinga W, van Ommen B, Rozing J (1984) Toxicity of Bis(tri-n-butyltin)oxide in the rat. II. Suppression of thymus dependent immune responses and of parameters of non-specific resistance after short-term exposure. Toxicol Appl Pharmacol 75:387-408

Vos JG, Van Logten MJ, Kreeftenberg JG, Kruizinga W (1984b) Effect of triphenyltin hydroxide on the immune system of the rat. Toxicology 29:325-336

Webber RJ, Dollins SC, Harris M, Hough AJ Jr (1985) Effects of alkyltins on rabbit articular and growth-plate chondrocytes in monolayer culture. J Toxicol Environ Health 16:229-242

Wong PTS, Chau YK, Kramar O, Bengert GA (1982) Structure-activity relationship of tin compounds on algae. Canad J Fish Aquat Sci 39:482-488

SELECTIVITY OF ANTIPROLIFERATIVE EFFECTS OF DIALKYLTIN COMPOUNDS IN VITRO AND IN VIVO

Gerhard Hennighausen, Institute of Pharmacology and Toxicology, Wilhelm-Pieck-University of Rostock, Leninallee 70, DDR-2500 Rostock, G.D.R.
Stanisław Szymaniec, Ludwik Hirszfeld Institute of Immunology and Experimental Therapy, Polish Academy of Sciences, ul. Czerska 12, 53-114 Wrocław, Poland

The inhibition of proliferation of thymocytes in vivo and in vitro is one of the earliest and most prominent biological effects caused by dialkyltin compounds. A single dose of dibutyltin dichloride (DBTC) or dioctyltin dichloride (DOTC) decreased the mitotic activity of thymocytes in mice and rats 1 to 7 days after administration. The inhibition of the mitotic rate of thymocytes was followed by a decrease of the thymus weight and thymocyte count. There was a difference in the time course of thymus atrophy induced by DOTC or glucocorticoids (Fig. 1). This difference could be explained by the additional cytolytic activity of glucocorticoids whereas the dialkyltin compounds induce the thymus atrophy solely by antiproliferative activity.

The thymocytes remaining in the thymus of mice and rats after treatment with dialkyltin dichlorides were characterized as mature thymocytes. These thymocytes showed high electrophoretic mobility and ecto-ATPase activity. It is supposed that DBTC and DOTC inhibit the proliferation of a rapid proliferating immature thymocyte subpopulation (Hennighausen et al 1985, Snoeij et al 1988).

The proliferation in vivo of cells other than thymocytes was also affected by dialkyltin compounds. In mice treated with DBTC, the proliferation of leukaemic cells L 1210 in the spleen was decreased. After single i.v. administration of DBTC or DOTC (4 and 8 mg/kg body weight) the number of leukaemic colony forming units in the spleen of mice was reduced to lower than 50% of controls. On the other hand, dimethyltin dichloride (DMTC) had no effect on proliferation of thymus cells in vivo and on the number of colony formed by leukaemic cells (Fig. 2).

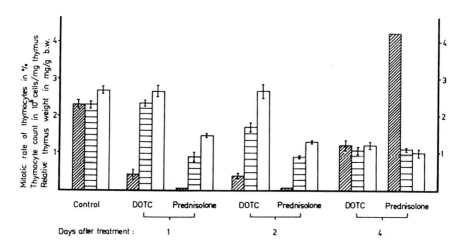

Fig. 1: Mitotic rate of thymocytes (colchicine-arrested metaphases during 3 h, diagonal hatched columns), thymocyte count (hatched columns) and relative thymus weight (open columns) in male CBA mice after single administration of DOTC (8 mg/kg i.v.) or prednisolone (125 mg/kg i.p.). Mean ± S.E., n=5 mice per group.

The proliferation of haematopoetic stem cells in a spleen colony assay in irradiated mice (CFU) wa used to test the antiproliferative activity of dialkyltin compounds on stem cells in vivo. In sublethally irradiated mice (700 R) endogenous colonies (colony forming units, CFU) are observed as a result proliferation of survived stem cells. The animals were treated with DOTC (single i.v. injection of 8 mg/kg body weight) before and after irradiation. The treatment 2 days after irradiation reduced number of CFU in the spleen as a result of antiproliferative activity (Fig. 3). An unexpected result was observed when the same dose of DOTC was given 1, 2 or 5 days before irradiation (Fig. 3). Mechanisms of this radioprotective effect of DOTC could be due to changes in the radiosensitivity of stem cells or of regulatory cells. Changes of the microenvironment in spleen could also affect the cell proliferation (CFU) in spleen.

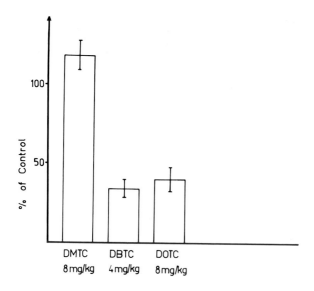

Fig. 2: Effects of dialkyltin dichlorides on the proliferation of leukaemic cells L 1210. Number of leukaemic spleen colony forming units in male mice (CD2F1) 7 days after i.v. injection of L 1210 cells and 6 days after treatment with organotins in percent of control. Mean ± S.E., n=9-19 mice per group.

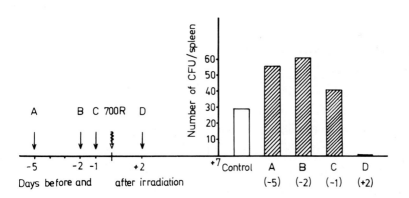

Fig. 3: Effects of DOTC on the proliferation of haematopoetic stem cells in sublethally irradiated male J129 mice. Left: Treatment with DOTC (8 mg/kg i.v.) before or after irradiation. Right: Number of endogeneous spleen colony forming units (CFU) 7 Days after irradiation. Mean of 15 mice per group.

The mitotic activity of cells in the bone marrow in vivo was not affected by dialkyltin compounds. A single dose of 8 mg/kg body weight of DOTC i.v. did not inhibit the mitotic rate of bone marrow cells in mice (Fig. 4).

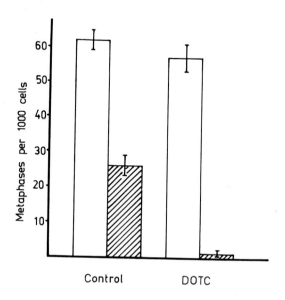

Fig. 4: Mitotic activity of bone marrow cells (open columns) and thymocytes (hatched columns) in male CBA mice 2 days after treatment with DOTC (8 mg/kg i.v.). Number of colchicine-arrested metaphases per 1000 cells 3 hours after administration of colchicine. Mean ± S.E., n=3-6 mice per group.

The proliferation of bone marrow stem cells was studied in a spleen colony assay in lethally irradiated recipient mice. When bone marrow cells from donor mice pretreated with 8 mg/kg body weight DOTC i.v. were transferred to lethally irradiated recipients the number of CFU in spleen did not differ as compared to controls (Table 1). This result is in agreement with findings in DBTC-treated mice (Penninks et al 1985). But when the bone marrow cells are transferred to irradiated animals and growth as CFU in the spleen of DOTC-treated mice was monitored than the antiproliferative effect of DOTC could be demonstrated (Table 2). Different pharmacokinetic conditions may be an important reason for these diverse effects.

Table 1: The potency of bone marrow cells (BMC) from DOTC-treated donor J 129 mice to form colony forming units (CFU) in the spleen of lethally irradiated recipient J 129 mice

Donors of BMC	Irradiated recipients of BMC	
	Number of injected cells bone marrow cells	Number of CFU/spleen[1] 7 days after 900 R and BMC injection
Control	0	0.3
Control	1×10^5	31.2
DOTC (8 mg/kg i.v.) on day - 2	1×10^5	33.6

[1] mean of 15 male mice per group

Table 2: The effect of DOTC (8 mg/kg i.v.) on proliferating bone marrow cells (BMC) as colony forming units (CFU) in the spleen of irradiated recipient CBA mice

Treatment of recipient	Number of injected BMC	Number of CFU/spleen[1] 7 days after 900 R and BMC injection
Control	1×10^5	25.2 ± 2.3
DOTC 24 h after BMC	1×10^5	3.6 ± 2.5
DOTC 48 h after BMC	1×10^5	9.8 ± 3.1

[1] mean $\pm$ S.E., n=20 male mice per group

In rats the number of nucleated bone marrow cells was not decreased but increased 5 days after oral administration of 18 mg/kg DBTC. A single dose of 9 or 18 mg/kg body weight DBTC (per os) diminished the acute depressive effects of benzene on the bone marrow in rats.

In vitro the mitotic activity of mouse and rat thymocytes was inhibited by DBTC and DOTC at concentrations of 10^{-8} to 10^{-6} mol/l (Fig. 5). The mitotic rate of thymocytes was more than 100 times more sensitive to the alkyltins than the viability test (Hennighausen et al 1985). The antimitotic effect was not reversible by washing 3 times of mouse thymocytes after an incubation time of 20 min with DBTC or 60 min with DOTC (Table 3).

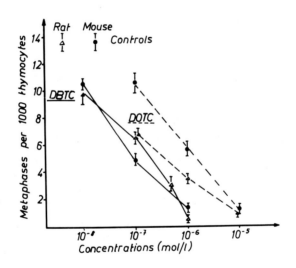

Fig. 5: Concentration-dependent inhibition of the mitotic activity (colchicine arrested metaphases) of thymocytes in vitro by DBTC and DOTC. Samples of 5×10^6 cells were incubated at 37°C in Eagle-MEM supplemented with 10 % fetal calf serum and 0.02M HEPES-buffer. Organotins were added at the beginning of incubation; after 2 hours colchicine was added and the incubation was continued for additional 3 hours. Mean $\pm$ S.E., n=15-20 samples per group.

Table 3: The mitotic activity of CBA mouse thymocytes after pretreatment in vitro with DBTC and DOTC[1]

Preincubation time with DBTC or DOTC (min)	Number of metaphases per 1000 cells after pretreatment with	
	DOTC 10^5 mol/l	DBTC 10^6 mol/l
0	13.5 ± 0.33	12.7 ± 0.44
20	13.0 ± 1.43	4.4 ± 0.49
60	7.0 ± 0.48	
120	4.9 ± 0.86	
300	2.2 ± 0.49	1.5 ± 0.19

1) Cells were preincubated with DBTC or DOTC, washed 3 times and incubated with colchicine for 180 min. Mean ± S.E., n=5 per group

Table 4: Cytotoxicity of DBTC on thymocytes of AKR mice and on AKS L4 leukaemic cells in vitro[1]

Cells	Concentrations of DBTC in mol/l		
	2.5×10^{-5}	1.25×10^{-5}	2.5×10^{-6}
	% of dead cells after incubation		
Thymocytes	100	98.3	50.0
AKS L 4 Leukaemic cells	23.4	20.5	14.5

1) Cells were incubated with DBTC for 1 h at 37°C and trypan blue was used for cell viability test. AKS leukaemic cells were from AKR mice injected ip with this cells 3 days earlier. Cells from peritoneal cavity were washed 2 times before using in cytotoxic test with DBTC.

Compared with mouse thymocytes, different leukaemic cells were less sensitive to cytotoxic effect of DBTC (Table 4). The spontaneous migration of thymocytes and macrophages was inhibited in vitro by the same concentration of DOTC or DBTC (Hennighausen and Lange 1980). The pharmacokinetics of dialkyltin compounds may influence the selectivity of the cytotoxic effects and of the cytostatic effect in highly proliferating cells. The concentrations of organotin determined in the thymus at 1 to 5 days after treatment of rats with DOTC or DBTC by atomic absorption spectrophotometry were much lower than in liver and spleen (Hennighausen et al 1980, 1982). But the organotin concentrations measured in the thymus are high enough to inhibit the mitotic activity of thymocytes (Table 5). Preliminary studies indicate very low concentrations of organotin in the bone marrow of rats treated with DOTC.

Table 5: Concentrations of organotin [1] in rat thymus 1-4 days after single i.v. administration of 8 mg/kg DBTC or DOTC and antimitotic concentrations in vitro

Compounds	Concentrations ($\mu mol/l$) or organotin in rat thymus 1-4 days after treatment	Concentrations ($\mu mol/l$) of DBTC and DOTC inhibiting the mitotic activity of rat thymocytes in vitro	
		50% inhibition	95% inhib.
DBTC	10 - 20	0.07	1
DOTC	5.8 - 6.2	0.1	10

1) Organotin was extracted with benzene or toluene and determined by measuring the tin using an AAS 1200 Varian, CRA63; electrothermal determination at 286.2 nm

The selectivity of cytotoxic and cytostatic effects of dialkyltin compounds in different cell types may be related to the capacity of cytoplasmatic proteins to bind organotin and thus prevent its reaction with vital cellular targets. Studies of the subcellular distribution have shown a higher content of DBTC in the cytoplasm of hepatocytes compared with thymocytes and a possible role of metallothioneins as scavanger has been suggested (Penninks and Seinen 1983).

An other reason for the different intracellular distribution of organotin in hepatocytes and thymocytes could be the higher content of cytosolic glutathione S-transferases in hepatocytes (Hennighausen 1986). These proteins are able to bind dialkyltin compounds by hydrophobic bonding at non catalytic binding sites (Hennighausen and Merkord 1985). The relative and absolute low capacity of thymocytes to detoxify organotins by binding on glutathione S-transferases, metallothioneins or other nonvital targets in the cytoplasm may increase the sensitivity of thymocytes to these compounds.

References

Hennighausen g, Lange P, Stülpner H, Ambrosius H, Karnstedt U (1980) Über Wirkungen von Dialkylzinnsalzen auf das Immunsystem. Acta biol med germ 39:149-155

Hennighausen G, Lange P (1980) A simple technique of testing for the influence of metal salts and other chemicals on macrophages and thymocytes in vitro. Arch Toxicol Suppl 4:143-147

Hennighausen G, Karnstedt U, Lange P (1981) Organotin concentrations in liver, spleen and thymus of rats after a single administration of di-n-octyltin dichloride. Pharmazie 36:710-711

Hennighausen G, Claus R, Rychly J, Schütt W (1985a) Electrophoretic mobility, ecto-ATPase activity and mitotic rate of thymocytes after treatment of mice with immunosuppressive drugs. In: Schütt W, Klinkmann H (ed) Cell electrophoresis. Walter de Gruyter, Berlin New York, p 657

Hennighausen G, Rychly J, Szymaniec S (1985b) On the sensitivity of several indicators of cytotoxic drug effects on thymocytes in vitro. In: Schütt W, Klinkmann H (ed) Cell electrophoresis. Walter de Gruyter, Berlin New York, p.663

Hennighausen G, Merkord J (1985) Meso-2,3-dimercaptosuccinic acid increases the inhibition of glutathione S-transferase activity from rat liver cytosol supernatants by di-n-butyltin dichloride. Arch Toxicol 57:67-68

Hennighausen G, (1986) Effect of diethyl maleate on non-protein sulfhydryl content and cellular functions of mouse thymocytes in vitro. Arch Toxicol Suppl 9:297-301

Penninks A H, Seinen W (1983) The lymphocyte as a target of toxicity: a biochemical approach to dialkyltin induced immunosuppression. In: Hadden J W, Chedid L, Dukor P, Spreafico F, Willoughby D (ed) Advances in Immunopharmacology vol II Pergamon Press Oxford New York, p 41

Penninks A H, Kuper F, Spit B J, Seinen W (1985) On the mechanism of dialkyltin-induced thymus involution. Immunopharmacology 10:1-10

Penninks A H, Seinen W (1987) Immunotoxicity of organotin compounds. A cell biological approach to dialkyltin induced thymus atrophy. In: Berlin A, Dean J, Draper M H, Smith E M B, Spreafico F (ed) Immunotoxicology Martius Nijhoff Publisher Dodarecht Boston Lancaster, p 258

Snoeij N J, Penninks A H, Seinen W (1988) Dibutyltin and tributyltin compounds induce thymus atrophy in rats due to a selective action on thymic lymphoblasts. Int J Immunopharmacol 10:891-899

COMPUTER ASSISTED STRUCTURE-ANTILEUKEMIC ACTIVITY CORRELATIONS OF ORGANOTIN COMPOUNDS AND INITIAL EXPLORATION OF THEIR POTENTIAL ANTI-HIV ACTIVITY

V. L. Narayanan, M. Nasr[*] and K.D. Paull
National Institutes of Health
NCI and NIAID[*]
Bethesda, Maryland 20892
U.S.A.

INTRODUCTION

The National Cancer Institute (NCI) and others have a continuing interest in exploring the anticancer and recently the anti-HIV potential of metal and metalloid compounds (Cleare et al 1980; Lippard 1983; Crowe et al 1980). The number of metal-containing compounds in the NCI collection is given in Table 1. Organotin compounds are the largest class among metals as represented by more than 2,000 compounds. This emphasis is the natural consequence of the wide biological use (Cardarelli 1972; Arakawa et al 1988) of tin compounds and their subsequent availability for screening by the NCI. A good deal of work has been done on the chemistry (Davis et al 1980; Wardell 1967), toxicology (Cardarelli 1986; Smith et al 1975), and metabolism (Thayer 1974; Kimmel 1977) of tin compounds. This study focuses on the structure-anticancer activity of tin-containing compounds evaluated by the NCI.

We report here an analysis of the various structural types of tin compounds tested by the NCI and the performance of these types as a group against the two *in vivo* murine leukemias P388 and L1210. This study seeks to utilize the computer's ability to group compounds according to precise structural definitions and to manipulate these groups using Boolean logic to create additional interesting subgroups. This technique has been used by us previously to analyze other classes of compounds (Nasr et al 1984; Paull et al 1984; Sadler et al 1984).

RESULTS

Many tin compounds are active against the standard P388

Table 1: Metal and Metalloid Compounds Tested by the National Cancer Institute

Transition elements										III A	IV A	V A	VI A
										Al 90			
Sc 8	Ti 87	V 84	Cr 253	Mn 273	Fe 883	Co 844	Ni 734	Cu 1275	Zn 861	Ga 59	Ge 226	As 1254	Se 734
Y 10	Zr 59	Nb 28	Mo 170	Tc 0	Ru 145	Rh 257	Pd 435	Ag 145	Cd 124	In 18	Sn 2008	Sb 353	Te 51
La+ 27	Hf 15	Ta 15	W 118	Re 41	Os 25	Ir 41	Pt 1538	Au 121	Hg 823	Tl 37	Pb 288	Bi 81	Po 0
Ce 30	Pr 23	Nd 27	Pm 0	Sm 20	Eu 79	Gd 18	Tb 40	Dy 22	Ho 12	Er 18	Tm 12	Yb 13	Lu 11

202

leukemia test system routinely used by the NCI, but only one of 700 tested showed confirmed activity (Paull et al 1986) against the standard L1210 leukemia system. Thus, tin compounds as a group exhibit a high frequency of activity against P388, and a low frequency of activity against L1210, although individual compounds may or may not be highly active against either system. This usage of "high activity" meaning a high frequency of actives instead of a high level of activity, is used throughout this study.

Approximately 29% of all tin compounds have demonstrated confirmed activity against P388, but only 1% against L1210. These percentages are slightly higher than calculated based on the numbers of confirmed actives in Table 2. The reason for the difference has been described (Nasr et al 1984; Paull et al 1984). The percentages used throughout Table 2 are computed using a projection technique. This technique uses historical data to estimate the expected outcome of partially completed testing on a set of compounds.

To understand the relevance of these findings, it is essential to compare these results with NCI's total screening experience against both P388 and L1210 systems comprising approximately 150,000 compounds in each category. Using the projection technique described above, we find the overall percentage of actives for P388 and L1210 to be 7.6% and 1.8% respectively. Using these overall percentages, one finds that tin compounds have been active against P388 almost six times as frequently as the overall percentage would predict and about the same frequency of activity against L1210 as one would have expected. Thus, P388 is far more sensitive than expected to tin compounds while L1210 is not especially sensitive. The reason for this differential sensitivity is not known. As will be noted below, this unpredicted sensitivity of P388 to tin compounds becomes far more evident once the structural details of the tin compounds are considered.

DISCUSSION

Table 2 shows the basic structural types of tin compounds and shows the effect of structural modifications on antitumor activity. The total number of compounds containing the indicated substructure that have been tested in any _in vivo_ or _in vitro_ test system is

Table 2

Z code no.	Substructure[a]	No. of compounds	P388/L1210 No. tested	No. CA[b]	% Activity	NSC[c]	Examples Structure
1	All Sn	2010	973 / 700	259 / 1	29 / 1	329882	$H_2N-CH_2-CH_2-S-Sn-Me$ with Cl, Me substituents
2	4-Coordinate $R_2Sn(HT)_2$	450	228 / 147	118 / 1	54 / 1	8786	Bu_2SnOAc_2
3	R_2SnO_2	200	90	52	60	345311	$Ph_2Sn-O-SnPh_2$ with ONO_2 groups
4	Me_2SnO_2	31	21	7	33	348067	$Me_2Sn-O-SnMe_2$ with OCOEt groups
5	Et_2SnO_2	25	17	16	94	221283	$(CH_2-SO_3)_2Sn-Et_2$ camphor derivative; Active P388 & L1210
6	Bu_2SnO_2	93	35	19	59	146118	Ph-substituted dioxastannolane with $SnBu_2$

Table 2 (continued)

Z code no.	Substructure[a]	No. of compounds	P388/L1210 No. tested	No. CA[b]	% Activity	NSC[c]	Examples Structure
7	Ph_2SnO_2	23	9	8	89	356206	$Ph_2Sn(OCOPh)_2$
8	$R_2Sn(OCO)_2$	56	26	10	40	306914	![structure: cyclic diacetate with SnEt2, C(O)O bridges]
9	$(R_2Sn)_2\text{–}O$ (with 1,0)	7	3	3	100	294263	MeOCO–Sn(Et)$_2$–O–Sn(Et)$_2$–OCOMe
10	$R_2Sn(OSO_2C)_2$	9	4	3	75	306907	$Et_2Sn(OSO_2Me)_2$
11	R_2SnS_2	85	42	18	46	202858	$Me(CH_2)_3\text{-}Sn(SCN)_2(CH_2)_3Me$
12	Me_2SnS_2	31	18	7	39	351601	$Me_2Sn(SP(S)Me_2)_2$
13	Bu_2SnS_2	28	13	4	41	202886	$Bu_2Sn(SBu)_2$

Table 2 (continued)

Z code no.	Substructure[a]	No. of compounds	P388/L1210 No. tested	P388/L1210 No. CA[b]	P388/L1210 % Activity	Examples NSC[c]	Examples Structure
14	Ph_2SnS_2	12	4	4	100	351602	$Ph_2Sn(SP(S)Ph_2)_2$
15	R_2SnN_2	28	25	11	48	292415	$Me_2Sn(NCS)_2$
16	R_2SnN_2 (N_2 are ring nitrogens)	18	15	9	61	342927	(imidazolyl)$_2$SnMe$_2$
17	R_2SnX_2	74	38	14	39	302600	Et_2SnCl_2
18	R_3Sn-HT	415	171 / 222	16 / 0	14 / 0	358324	1,4-C$_6$H$_4$(SnEt$_2$Cl)$_2$
19	R_3Sn-O	196	80	3	10	173032	$(CH_2=CH)_3-Sn-OAc$
20	R_3Sn-S	76	34	3	11	161513	$Ph_3Sn-S-C(S)$-N(morpholinyl)

Table 2 (continued)

Z code no.	Substructure[a]	No. of compounds	P388/L1210 No. tested	P388/L1210 No. CA[b]	P388/L1210 % Activity	NSC[c]	Examples Structure
21	R_3Sn-N	44	23	0	3	142120	Me-C6H4-SO2NHSnEt3
22	R_3Sn-X	78	29	10	39	202912	Et3SnF
23	R_3Sn-X (two moieties in same molecule)	5	4	2	50	341996	Me2(Cl)Sn-C6H4-Sn(Cl)Me2
24	RSn-(HT)$_3$	38	15	3	20	294232	Ph-Sn(Cl)-O-(CH2)2-NH-CH2-CH2-O
25	RSnO$_2$Cl	6	6	3	50	294232	
26	Sn(HT)$_4$	142	56	1	5	229572	Sn(O,S cage)

Table 2 (continued)

Z code no.	Substructure[a]	No. of compounds	P388/L1210 No. tested	No. CA[b]	% Activity	NSC[c]	Examples Structure
27	R_4Sn	368	188 149	8 0	5 0	348111	$(H_2C=HC-CH_2)_2 SnEt_2$
28	R_2Sn | $(CH_2COOMe)_2$	5	5	4	80	323991	$Et_2Sn(CH_2COOMe)_2$
29	All 5-coordinate Sn compounds	137	73 68	11 0	19 1	162794	
30	$R_2Sn(HT)_3$	18	16	10	68	162794	
31	All 6-coordinate Sn compounds	326	230 104	83 0	39 1	303784	
32	$R_2Sn(HT)_4$	172	163	79	50	303784	

Table 2 (continued)

Z code no.	Substructure[a]	No. of compounds	P388/L1210			NSC[c]	Examples Structure
			No. tested	No. CA[b]	% Activity		
33	$R_2SnX_2N_2$	116	113	55	49	292424	
34	$R_2SnO_2N_2$	10	10	6	80	292424	
35	R_2SnN_4	14	14	8	58	326390	
36	$R_2SnN_2S_2$	5	5	2	40	334724	
37	R_2SnO_4	12	12	4	41	254041	

Table 2 (continued)

Z code no.	Substructure[a]	No. of compounds	P388/L1210 No. tested	No. CA[b]	% Activity	NSC[c]	Examples Structure
38	$R_2SnO_2S_2$	4	4	1	25	297324	
39	Y_2SnX_4 (Y = any atom)	36	14	2	15	157846	
40	N_2SnX_4	8	8	2	26	7890	
41	$R_3Sn(HT)_3$	6	4	0	5	297500	

Table 2 (continued)

Z code no.	Substructure[a]	No. of compounds	P388/L1210			Examples	
			No. tested	No. CA[b]	% Activity	NSC[c]	Structure
42	Sn(HT)$_6$	132	47	3	12	320529	bipyridyl-SnCl$_4^{+4}$
43	SnX$_6$	19	3	0	3	43602	SnCl$_6^{+4}$ 2 NH$_4^+$
44	[R$_2$Sn]x	6	4	2	50	162819	[Ph$_2$Sn]$_x$
45	[(R$_2$Sn)O]x	27	15	6	44	323987	[(PhBuSn)O]$_x$
46	[(R$_2$Sn)S]x	8	4	1	25	92620	[(Bu$_2$Sn)S]$_x$
47	(HT)$_2$Sn-R	10	4	2	50	96391	Me(CH$_2$)$_3$SnO$_2$H
48	Sn-N-CCOO (amino acids)	23	21	7	37	346395	Sn bis(phenylalaninate) chelate

Table 2 (continued)

Z code no.	Substructure[a]	No. of compounds	P388/L1210 No. tested	No. CA[b]	% Activity	NSC[c]	Examples Structure
49	(structure with Sn, N, O, C=O ring)	17	16	7	48	358913	$\overline{Bu}_2Sn$–NH$_2$ with Me, Me, S, +4 (structure)
50	(bicyclic structure with H$_2$N, Sn, N, O, C=O)	4	4	4	100	326392	(Sn+2 complex with Bu, Bu, NH$_2$, C=O groups)
51	Sn–O–S(O)R	15	14	0	5	251424	Bu$_2$Sn(OS(O)Ph)$_2$
52	Sn — Sn	18	3	0	7	92633	n-Bu$_3$Sn-SnBu$_3$-n
53	Sn — M M = any metal other than Sn	16	8	3	38	294516	(Ph$_3$Ge – Sn(O)Et)$_3$

[a] HT = any atom except C or H; X = halogen.
[b] Confirmed actives, i.e., compound showed activity in two separate tests.
[c] The NSC identify each compound in the NCI collection.

shown. In addition, Table 2 depicts the number of compounds that have been tested against P388 and/or L1210 leukemia, the number of confirmed actives (C.A.) and the percentage of activity (Nasr et al 1984; Paull et al 1984). For each substructure type, an example of an active compound is given except in cases where no active was found.

The information generated in Table 2 is indicative rather than definitive. The presence of relatively many active compounds within an analyzed group is considered a reasonable basis for additional, in depth studies on the group. It should be noted that a high percentage of actives cannot be considered proof that the subject substructure is required for the antitumor activity or even relevant to it. On the other hand, if few active compounds are found among a relatively large group having a given substructure, it is safe to assume that the particular substructure is not particularly relevant to that type of anticancer activity.

The results of our analysis are discussed below.

I. 4-Coordinate Tin Compounds

    A. Diorganotins $R2Sn-(HT)_2$

        This represents a highly active class against P388. The activity varies somewhat with the nature of the organic substituent R and the electronegative substituent HT. In general, the diorganotin compounds (Z2) showed 54% activity against P388; 60% with HT=O (Z3), 46% with HT=S (Z11), 48% with HT=N (Z15), and 39% with HT=halogen (Z17). Excellent activity of 94% against P388 was obtained with R=Et, HT=O (Z5), and 89% with R=Ph, HT=O (Z7). All the three bis-diorganotin compounds (Z9) tested against P388 demonstrated confirmed activity. When R=Ph, HT=S (Z14), the four compounds tested against P388 are all confirmed actives. The diorganotin nitrogen compounds with the nitrogens as part of a ring system (Z16) showed 61% activity against P388.

    B. Triorganotin Compounds R_3Sn-HT

        The class (Z18) showed 14% activity against P388 and no activity against L1210. A very advantageous electronegative substituent in this class appears to be halogen (Z22); and within that class, the subclass (Z23)

C. Monoorganotin Compounds $RSn-(HT)_3$

The overall activity, 20%, for this category (Z24) is derived from the activity, 50%, of the subtype $RSnO_2Cl$ (Z25). No other compound of the $RSn-(HT)_3$ class showed any activity against either P388 or L1210.

D. Tetraorganotin Compounds R_4Sn

This class (Z27) in general showed poor activity, but the subclass (Z28) showed 80% activity. The active compounds possess two carboxyalkyl ligands. It has been postulated that the cleavage of a carboxyalkyl ligand to tin is enhanced by the ability of the carboxy group to stabilize the carbanion - CH_2COOR in the transition state (Wardell 1967). The antileukemic activity of these compounds might be related to the ease of cleavage of such ligands.

E. Tetraheterotin Compounds $Sn-(HT)_4$

This class showed 5% activity against P388 (Z26) and as such, represents the least active class of tetracoordinated tin compounds. Compounds with two nitrogen ligands and either two halogen, or two oxygen ligands were devoid of any P388 activity.

II. 5-Coordinate Tin Compounds

As a class, 5-coordinate tin compounds have good P388 activity, 19% (Z29); the diorganotin type (Z30) is the only significantly active, subclass (68%).

III. 6-Coordinate Tin Compounds

A. Diorganotin Compounds $R_2Sn-(HT)_4$

Most of these show good activity against P388. The best subclass (Z34) has two oxygen and two nitrogen ligands. No diorganotin compound with four halogen ligands has been tested.

B. Triorganotin Compounds $R_3Sn-(HT)_3$

Only four compounds (Z41) were tested against P388 and they are inactive.

C. Hexaheterotin Compounds $Sn-(HT)_6$

This class showed 12% activity against P388 (Z42).

Compounds with 2 nitrogens and 4 halogens (Z40) showed 26% activity against P388.

IV. Miscellaneous Types
   A. Diorganotin Compounds R_2Sn
      Of the four compounds (Z44) tested against P388, two are active.
   B. Diorganotin Compounds $[(R_2Sn)HT]_x$
      Compounds of type $[(R_2Sn)O]_x$ (Z45) showed 44% activity against P388 and 9% activity against L1210. (Among the 8 tested, one compound showed initial activity against L1210). Of the four compounds of type $[(R_2Sn)S]_x$ (Z46), that were tested against P388, one showed confirmed activity. Of the two compounds tested against L1210, one showed initial activity.
   C. Monoorganotin Compounds $RSn-(HT)_2$
      Of the four monoorgano compounds (Z47) tested against P388, two have shown confirmed activity.
   D. Tin Aminoacids
      Tin aminoacids showed 37% activity against P388 (Z48). All four cyclic tin aminoacids (Z49) are active against P388.
   E. Tin Sulfinates
      Of the fourteen compounds that were tested against P388, only 5% are active (Z50).
   F. Compounds with Tin Bonded to Tin or to a Different Metal
      The three compounds with the tin bonded to another tin (Z51), were all inactive against P388. In contrast, compounds with tin bonded to different metals (Z52), showed good activity against P388.

<u>Other areas of biological interest</u>:

Recently, Ward et al (1988) has reported the antiherpes activity of some antitumor organotin compounds. This finding has prompted us to explore the anti-HIV activity of a selected group of organotin compounds using the newly developed microculture antiviral assay (Weislow et al 1989) that measures the ability of compounds to reverse the HIV-induced viral cytopathic effects against CEM-SS cell lines. So far compounds belonging to two structural types, one previously studied for antiherpes activity (Ward et al 1988) and the other for suppression of cell proliferation (Arakawa et al 1988) have not shown anti-AIDS activity in our screening system (Weislow et al 1989). It is hoped that the availability of this assay will facilitate the prompt identification and development of unique organotin compounds as potential anti-HIV drugs.

Morgan et al (1987) have shown that tin (IV) etiopurpurin dichloride ($SnEt_2$), NSC 619679, to be a photosensitizer useful in photodynamic therapy. Its ease of synthesis, purity and *in vitro* and *in vivo* antitumor effects warrant its further investigation as an alternate drug to DHE, a comple mixture of porphyrins which is currently the drug of choice for photodynamic therapy. Therefore, NCI is actively pursuing this interesting lead.

CONCLUSIONS:

Many organotin compounds have demonstrated anticancer activity against the murine P388 leukemia which was utilized routinely up until recently as the primary screen by the National Cancer Institute. The diorganotin compounds, regardless of their overall coordination number, constitute the most frequently active subclass. In contrast, organotin compounds are generally inactive against the L1210 leukemia. Also, it should be noted that a selected group of 24 P388 active tin compounds was further tested in the 5 murine and 3 human tumor xenograft test systems of the NCI tumor panel (Venditti et al 1984). None of them showed activity that would warrant further development. Recently, we have begun to investigate the anti-HIV activity of tin compounds using the newly developed assay system. Tin (IV) etiopurpurin dichloride is being developed for potential use in photodynamic therapy.

REFERENCES

Arakawa Y, Wada O (1988) Suppression of cell proliferation by certain organotin compounds. In: Zuckerman JJ (ed) Tin and malignant cell growth, CRC Press, Boca Raton FL, p 83

Cardarelli NF (1972) Slow release pesticides utilizing organotins. Tin and its uses vol 93 p 16

Cardarelli NF (ed) (1986) Tin as a vital nutrient: implications in cancer prophylaxis and other physiological processes. CRC Press, Inc., Boca Raton FL

Cleare MJ, Hydes PC (1980) Antitumor properties of metal complexes. Metal ions in biological systems vol 11 p 1
Crowe AJ, Smith PJ, Atassi G (1980) Investigations into the antitumor activity of organotin compounds. Chem Biol Interactions 32:171
Davis AG, Smith PJ (1980) Recent advances in organotin chemistry. Adv. Inorg Chem Radiochem 23:41
Kimmel EC, Fish RH, Cadisa JE (1977) Metabolism of organotin compounds in microsomal monooxygenase systems and in mammals. J Agric Food Chem 25(1):1-9
Lippard SJ (ed) (1983) Platinum, gold, and other chemotherapeutic agents. ACS Symposium Series. Washington, DC, p 209
Morgan AR, Garbo GM, Keck RW, Selman SH (1987) Tin (IV) etiopurpurin dichloride: an alternative to DHE? New Directions in Photodynamic Therapy 847:172-179
Narayanan VL (1983) Strategy for the discovery and development of novel anticancer agents. In: Reinhoudt DN, Connors TA, Pinedo HM, van dePall KW (eds) Developments in pharmacology, vol 3. Martinus Nijhoff, The Hague, Boston, p 5
Nasr M, Paull KD, Narayanan VL (1984) Computer structure activity correlations. Adv Pharmacol Chemother, vol 20 p 123
Paull KD, Nasr M, Narayanan VL (1984) Computer assisted structure activity correlations of benzo (de) isoquinoline 1,3-diones. Arzneim-Forsch/Drug Research 34(II):1243-1246
Paull KD, Hodes L, Simon RM (1986) Efficiency of antitumor screening relative to activity criteria. JNCI 76(6):1137-1142
Rosenberg B (1980) Clinical aspects of platinum anticancer drugs. Metal Ions in Biol Syst vol 11 p 168
Sadler PJ, Nasr M, Narayanan VL (1984) The design of metal complexes as anticancer drugs. Dev Oncol 17:290-304
Smith PJ, Smith L (1975) Organotin compounds and applications. Chem Br 11(6):208-212
Thayer JS (1974) Organometallic compounds and living organisms. J Organomet 76(3):265-295
Venditti JM, Wesley RA, Plowman J (1984) Current NCI preclinical antitumor screening in vivo. Adv Pharmacol Chemother, vol 20 p 1
Ward SG, Tylor RC, Crowe AG (1988) The in vitro antiherpes activity of some selected antitumor organotin compounds. Appl Organomet Chem 2:47-52
Wardell JL (1967) Reactions of electorphilic reagents with tin compounds containing organofunctional groups. In: Zuckerman JJ (ed) Organotin compounds: new chemistry and applications ACS Advances in Chemistry Series 157, Washington, DC, p 113
Weislow OS, Kiser R, Fine DL, Bader J, Shoemaker RH, Boyd MR (1989) New soluble-formazan assay for HIV-1 cytopathic effects. JNCI 81:577-586

ROUTE OF ADMINISTRATION IS A DETERMINANT OF THE TISSUE DISPOSITION AND EFFECTS OF TBTO ON CYTOCHROME P-450-DEPENDENT DRUG METABOLISM

Daniel W. Rosenberg
The Rockefeller University Hospital
1230 York Avenue
New York, New York, 10021 USA.

INTRODUCTION

Organotins have found wide industrial and agricultural applications as a result of the potent biocidal activities exhibited by many of these compounds (1,2). Recently, some of these compounds have undergone testing for potential antitumor activity (3). This wide diversity of applications of organotins has prompted an extensive evaluation of their general toxicity (4-7).

Despite an existing literature on the biotransformation of organotin compounds via the hepatic cytochrome P-450-dependent monooxygenase system (8), the specific metabolic products that are responsible for many of these biological effects have not been determined. It is apparent, however, that side-chain hydroxylation and destannylation reactions can occur both in vitro and in vivo (8,9) in both the liver and small intestinal epithelium.

In the present studies, we have examined the ability of bis(tri-n-butyltin)oxide (TBTO), a trialkyltin compound with numerous agricultural applications (10), to produce alterations in heme oxygenase, the rate-limiting enzyme in the oxidative catabolism of heme to bile pigments (11). Concomitant changes in cytochrome P-450 content and functional activity in the liver and epithelial cells of the proximal small intestine are described as well.

MATERIALS AND METHODS

Materials. TBTO was generously provided by the M & T Chemical Company (Rahway, NJ), and was at least 95% pure as determined by the manufacturer and confirmed by thin-layer chromatography. Sodium heparin (1,000 Units/ml) was purchased from Riker Laboratories Inc. (Northridge, CA). All other reagents used in these studies were purchased from Sigma Chemical Co. (St. Louis, MO).

Treatment of animals. Male Sprague-Dawley rats (200-250 g) were purchased from Taconic Farms (Germantown, NY). Rats were maintained on Standard Purina Rodent Laboratory Chow 5001 (St. Louis, MO) and were allowed to acclimatize to a light-cycled room (12 hours light/dark) for at least one week prior to study. TBTO was either dissolved in ethanol and administered subcutaneously (50 µmoles/kg body wt) in a single dose or suspended in corn oil and administered by gavage (50 µmoles/kg) in a single dose at the times indicated in the legends to Tables and Figures. Control animals received an equivalent volume (1.0 ml/kg) of either corn oil or ethanol.

Preparation of subcellular fractions and enzyme assays. Animals were sacrificed and then exhaustively perfused in situ with ice-cold 0.9% NaCl through the left ventricle. Liver microsomes were prepared as described earlier (12). The small intestine was cut at the pyloric junction and the entire length was irrigated in situ with ice-cold 0.9% NaCl to remove intestinal contents. The first 35 cm of small intestine was then removed and irrigated once more with cold saline. The first 5 cm segment was discarded and the remaining excised tissue was then placed onto a watch glass maintained on ice and cut longitudinally to expose the mucosal surface. The mucosal cells were gently scraped off, weighed and placed immediately into a medium containing ice-cold potassium phosphate buffer (0.1 M, pH 7.4) containing sucrose (0.25 M), 20% glycerin (v/v), trypsin inhibitor (5 mg/ml) and heparin (3 U/ml), as described by Stohs et al. (13). Following homogenization with a tight-fitting Potter-Elvehjem Teflon-glass homogenizer, the samples were subjected to sonication at 4°C using 3 five-second pulses at 30W/min. Microsomes were then prepared by differential centrifugation and suspended in an appropriate volume of potassium phosphate buffer (0.1 M, pH 7.4) to a protein concentration of 5-10 mg/ml.

The activity of heme oxygenase (EC 1.14.99.3) was determined as previously described (14), using a substrate hemin concentration of 50 μM with intestinal and hepatic microsomes. Bilirubin formation was calculated by using an absorption coefficient of 40 mM-1.cm-1 between 464 and 530 nm (15). Cytochrome P-450 content was measured in liver microsomes by the method of Omura and Sato (16). Intestinal cytochrome P-450 was measured in microsomal suspensions from the dithionite-reduced difference spectrum of CO-bubbled samples (17). Protein content was determined by the method of Lowry et al. (18), using bovine serum albumin as a standard.

Tin analysis. All tin analyses were performed by graphite furnace atomic absorption spectroscopy (GFAAS) on a Zeeman/5000 atomic absorption spectrophotometer equipped with an HGA-500 graphite furnace (Perkin-Elmer Corp., Norwalk, CT), using Zeeman background correction. The graphite furnace utilized the L'vov platform (19) with appropriate wavelength (286.3 nm) and instrumental operating conditions as recommended by the manufacturer and optimized in our laboratory for tin measurement. All tin analyses included magnesium nitrate (20μg) and ammonium phosphate (2 μg) in 20% nitric acid as a matrix modifier, added to the platform with the AS-40 autosampler. The furnace temperature parameters are shown in Table I.

Table I
Furnace conditions for tin analyses

	Step number						
	1	2	3	4	5	6	7
Temp, °C	150	250	1000	20	2100	2600	20
Ramp, sec.	1	15	10	1	0	1	1
Hold, sec.	30	0	20	15	5	5	15
Int. Gas Flow, ml/min.	300	300	300	300	0	300	300

Statistical analyses. Where appropriate, the data were analyzed by Dunnett's multiple comparison of treatments against a single control group and a P value <0.05 was regarded as statistically significant.

RESULTS AND DISCUSSION

Analysis of dosing suspensions. An analysis of TBTO dosing suspensions is shown in Table II. Tin content was determined by GFAAS. The mean tin concentration found in the corn oil dosing suspension was 96.3% of the target concentration, with a coefficient of variation (CV) of 6.6%. In our initial experiments, a dosing suspension containing 20% (v/v) ethanol was used. However, upon subsequent analysis, this suspension was found to be non-uniform. In all further experiments, TBTO was dissolved in 95% ethanol, from which tin recoveries ranged from 80-95% of the target with a CV of 6.6% (Table II).

Table II
Analysis of TBTO in Dosing Suspensions

Dosing Vehicle	TBTO Dose (mg/kg b.w.)	TBTO Target Dose (mg Sn/ml)	Mean Found (mg Sn/ml)	% of Target Concentration	CV, %
Corn Oil	29.80	11.87	11.43	96.3	± 11.6%
Ethanol, 20%	29.80	11.87	5.18	43.7	± 16.2%
Ethanol, 95%	29.80	11.87	10.22	86.1	± 6.6%

The effects of TBTO on organ and body weights. Following oral or parenteral treatment of rats with TBTO, food consumption and organ and body weights were measured. The effects of TBTO treatment at 48 hours on these parameters are shown in Table III. Average daily food consumption and body weight gain were substantially (50%) reduced in animals administered TBTO by either route. Oral treatment with TBTO produced a significant ($P<0.05$) increase ($\approx$30%) in epithelial cell weight from an initial mean weight of 1.69 ± 0.01 gm to a mean weight of 2.16 ± 0.07 gm. Despite the increase in epithelial cell weight, however, there were no observable changes in villous structure or underlying cell morphology in the proximal small intestine, at least at the light microscopic level (unpublished observations).

Table III
The acute effects of TBTO on organ and body weights at 48 hours

	Treatment Group		
	Control	Oral	SC
Initial Body Wt (g)	244 ± 5	287 ± 2	231 ± 3
Terminal Body Wt (g)	267 ± 6	280 ± 1	229 ± 3
Body Wt Gain (g)	22.7 ± 3.8	-6.7 ± 2.0	-2.0 ± 2.1
Ave. Daily Food Consumption (g/rat)	24.8	13.0	16.5
Liver Weight (g)	14.76 ± 0.58	13.43 ± 0.70	14.21 ± 0.49
Liver/Body Weight (%)	5.53 ± 0.15	4.79 ± 0.23	6.22 ± 0.20
Kidney Weight (g)	3.15 ± 0.23	2.67 ± 0.17	2.51 ± 0.11
Kidney/Body Weight (%)	1.18 ± 0.07	0.95 ± 0.06	1.10 ± 0.06
Epithelial Cell Weight (%)	1.69 ± 0.01	2.16 ± 0.07[a]	1.48 ± 0.11

TBTO was given either subcutaneously or by gavage in a single dose (50 μmoles/kg body wt). Rats were killed 48 hours later and body and organ weights measured. Values are reported as the means ± S.E.M. of three individual animals per group.

[a] These values were significantly different from the controls treated with vehicle alone ($p<0.05$, Dunnett's multiple comparison).

A comparison of TBTO effects on heme oxygenase, cytochrome P-450 and total tin in the liver and small intestinal epithelium. TBTO was given in a single dose (50 μmoles/kg body wt), either orally or subcutaneously, and the effects on heme oxygenase and cytochrome P-450 were examined 48 hours later in the liver and small intestinal epithelium. As shown in Table IV, TBTO administered subcutaneously produced substantial changes in heme catabolism in the liver, with heme oxygenase activity increased almost 3-fold above control levels, from a mean control value of 3.75 ± 0.25 nmoles bilirubin formed/mg protein per hr to a mean value of 9.99 ± 1.64 nmoles bilirubin formed/mg protein per hr. Cytochrome P-450 content was concomitantly reduced (60%) in parenterally exposed rats at this time point. Oral exposure to TBTO had no significant effects on either hepatic heme oxygenase activity or cytochrome P-450 content. Tin concentration, however, was markedly lower (80%) in the livers of rats parenterally exposed to TBTO (Table IV).

Table IV
TBTO effects on heme oxygenase and cytochrome P-450 in liver

Treatment	Heme Oxygenase (nmoles bilirubin/ mg per hour)	Cytochrome P-450 (nmoles/mg)	Tin Content (μg Sn/gm dry wt)
Controls	3.75 ± 0.24	0.66 ± 0.05	N.D.
TBTO, oral	4.40 ± 0.21	0.77 ± 0.07	20.18 ± 4.79
TBTO, sc	9.99 ± 1.64[a]	0.37 ± 0.03	2.29 ± 0.36

TBTO was given either subcutaneously or by gavage in a single dose (50 μmoles/kg body wt). Heme oxygenase, cytochrome P-450 and total tin were determined in liver microsomes 48 hours later as described under Materials and Methods. Values are reported as the means ± S.E.M of at least three individual animals per group.

N.D. Not detectable.

[a] $P<0.05$, Dunnett's multiple comparison of treatments against a single control.

In the small intestine, oral treatment with TBTO was required to produce induction of heme oxygenase activity at 48 hours (Table V), from a mean control value of 4.29 ± 0.26 nmoles bilirubin formed/mg protein per hr to a mean value of 6.37 ± 0.61. In addition, the content of cytochrome P-450 was reduced following oral treatment with the organotin (Table V).

Table V
TBTO effects on heme oxygenase and cytochrome P-450 in the small intestinal epithelium

Treatment	Heme Oxygenase (nmoles bilirubin/mg per hr)	Cytochrome P-450 (nmoles/mg)
Controls	4.29 ± 0.26	0.056 ± 0.009
TBTO, oral	6.37 ± 0.61	0.038 ± 0.008[a]
TBTO, sc	3.85 ± 1.08	0.050 ± 0.006

TBTO was given either subcutaneously or by gavage in a single dose (50 μmoles/kg body wt). Heme oxygenase activity and cytochrome P-450 content were determined in the small intestinal epithelium 48 hours later as described under Materials and Methods. Values are reported as the means ± S.E.M of at least three individual animals per group.

[a] $P<0.05$, Dunnett's multiple comparison of treatments against a single control.

Tin content at the subcutaneous site of injection. An explanation for the low concentration of tin found in the liver after parenteral treatment might reside in part in the slow systemic uptake of TBTO from the subcutaneous site of administration. Total tin recovered from the injection site was measured 48 hours after parenteral treatment with TBTO (50 μmol/kg body weight). This data is shown in Table VI.

Table VI
Tin content at the subcutaneous site of injection

Initial Body Wt (g)	Dose of TBTO Injected (mg/kg bw)	Total Tin Injected (mg)	Tin Recovered From Injection Site (mg)	% of Tin Recovered at Injection Site
231.0 ± 3.8	29.8	2.74 ± 0.04	1.60 ± 0.21	58.0 ± 6.6

TBTO was dissolved in 95% ethanol and injected subcutaneously in a single dose (50 μmol/kg body wt). Rats were killed 48 hours later and total tin remaining at the site of injection was determined by graphite furnace atomic absorption spectroscopy as described under Materials and Methods. Values are reported as the means ± S.E.M. of three individual animals.

In conclusion, the results of these studies demonstrate that TBTO interacts with hepatic and intestinal heme metabolism in a manner that is dependent on the route by which this compound is administered. The toxicity produced in the liver can be circumvented when TBTO is given by gavage, suggesting that intestinal "first pass" metabolism of this compound produces a metabolite that is less toxic to the liver. Furthermore, the site of action of TBTO can be shifted from the liver to the small intestinal epithelium when the compound is administered by gavage. This change in target tissue from liver to small intestine occurs despite a much higher content of tin accumulating in the liver after oral treatment. These observations raise the possibility that route of administration-dependent metabolite formation contributes to the biological tissue-specific effects that occur in rats in response to TBTO exposure.

ACKNOWLEDGEMENTS

The authors wish to thank Mr. Henry Roque, M.S. for his dedicated and highly skilled technical assistance. These studies were supported in part by USPHS grant ES-01055 and a generous gift from the Eugene and Theresa Lang Foundation. Computerized data collection and statistical analysis for this study was accomplished using the CLINFO system funded by the National Institutes of Health, General Clinical Research Center Grant M01 RR00102.

REFERENCES

1. WHO Environmental Health Criteria 15. 'Tin and organotin compounds.' WHO, Geneva 1980.

2. Gitlitz, MH. 'Organotins in agriculture.' In Zuckerman JJ, ed., Organotin Compounds: New Chemistry and Applications. American Chemical Society, 1976, Washington, DC, 167-176.

3. Zuckerman JJ (ed.). Tin and Malignant Cell Growth, CRC Press, 1988, Boca Raton, FL.

4. Reiter, L, Kidd, K, Heavner, G and Ruppert, P. 'Behavioral toxicity of acute and subacute exposure to triethyltin in the rat.' Neurotoxicology 2:97-112, 1980.

5. Aldridge, WN, Street, BW and Skilleter, DN. 'Oxidative phosphorylation. Halide-dependent and halide-independent effects of triorganotin and triorganolead compounds on mitochondrial functions.' Biochem. J. 168:353-364, 1977.

6. Seinen, W, Vos, JG, Brands, R and Hooykaas, H. 'Lymphocytotoxicity and immunosuppression by organotin compounds. Suppression of graft-versus-host reactivity, blast transformation, and E-rosette formation by di-n-butyltin dichloride and di-n-octyltin dichloride.' Immunopharmacology 1:343-355, 1979.

7. Magee, PN, Stoner, HR and Barnes, JM. 'The experimental production of oedema in the central nervous system of the rat by triethyltin compounds.' J. Path. Bact. 73:107-124, 1957.

8. Fish, RH, Kimmel, EC and Casida, JE. 'Bioorganotin chemistry: Reactions of tributyltin derivatives with a cytochrome P450 dependent monooxygenase enzyme system.' J. Organometallic Chem. 118:41-54, 1976.

9. Iwai, H, Wada, O, Arakawa, Y and Ono, T. 'Intestinal uptake site, enterohepatic circulation, and excretion of tetra- and trialkyltin compounds in mammals.' J. Toxicol. Environ. Health 9:41-49, 1982.

10. Piver, WT. 'Organotin compounds: Industrial applications and biological investigations.' Environ. Health Perspect. 4:61-79, 1973.

11. Tenhunen, R, Marver, MS and Schmid, R. 'The enzymatic conversion of heme to bilirubin by microsomal heme oxygenase.' Proc. Natl. Acad. Sci. USA 61:748-755, 1969.

12. Rosenberg, DW, Drummond, GS and Kappas, A. 'The influence of organometals on heme metabolism: In vivo and in vitro studies with organotins.' Mol. Pharmacol. 21:150-158, 1981.

13. Stohs, SJ, Grafstrom, RC, Burke, MD, Moldeus, PW and Orrenius, SG. 'The isolation of rat intestinal microsomes with stable cytochrome P-450 and their metabolism of benzo(a)pyrene.' Arch. Biochem. Biophys. 177:105-116, 1976.

14. Maines, MD, and Kappas, A. 'Cobalt stimulation of heme degradation in the liver.' J. Biol. Chem. 250:4171-4177, 1975.

15. Maines, MD and Kappas, A. 'Cobalt induction of hepatic heme oxygenase; with evidence that cytochrome P450 is not essential for this enzyme activity.' Proc. Natl. Acad. Aci. USA **71:** 4293-4297.

16. Omura, T and Sato, R. 'The carbon monoxide-binding pigment of liver microsomes. I. Evidence for its hemoprotein nature.' J. Biol. Chem. **239:**2370-2378, 1964.

17. Matsubara, T, Koike, M, Touchi, A, Tochino, Y and Sugeno, K. 'Quantitative determination of cytochrome P-450 in rat liver homogenate.' Anal. Biochem. **75:**596-603, 1976.

18. Lowry, OH, Rosebrough, NJ, Farr, AL and Randall, RJ. 'Protein measurement with the Folin phenol reagent.' J. Biol. Chem. **193:**265-275, 1951.

19. Slavin W, Manning DC, Carnrick GR. The stabilized temperature platform furnace. At. Spectrosc. **2:**137-145, 1981.

NATO ASI Series H

Vol. 1: Biology and Molecular Biology of Plant-Pathogen Interactions.
Edited by J. A. Bailey. 415 pages. 1986.

Vol. 2: Glial-Neuronal Communication in Development and Regeneration.
Edited by H. H. Althaus and W. Seifert. 865 pages. 1987.

Vol. 3: Nicotinic Acetylcholine Receptor: Structure and Function.
Edited by A. Maelicke. 489 pages. 1986.

Vol. 4: Recognition in Microbe-Plant Symbiotic and Pathogenic Interactions.
Edited by B. Lugtenberg. 449 pages. 1986.

Vol. 5: Mesenchymal-Epithelial Interactions in Neural Development.
Edited by J. R. Wolff, J. Sievers, and M. Berry. 428 pages. 1987.

Vol. 6: Molecular Mechanisms of Desensitization to Signal Molecules.
Edited by T. M. Konijn, P. J. M. Van Haastert, H. Van der Starre, H. Van der Wel, and M. D. Houslay. 336 pages. 1987.

Vol. 7: Gangliosides and Modulation of Neuronal Functions.
Edited by H. Rahmann. 647 pages. 1987.

Vol. 8: Molecular and Cellular Aspects of Erythropoietin and Erythropoiesis.
Edited by I. N. Rich. 460 pages. 1987.

Vol. 9: Modification of Cell to Cell Signals During Normal and Pathological Aging.
Edited by S. Govoni and F. Battaini. 297 pages. 1987.

Vol. 10: Plant Hormone Receptors. Edited by D. Klämbt. 319 pages. 1987.

Vol. 11: Host-Parasite Cellular and Molecular Interactions in Protozoal Infections.
Edited by K.-P. Chang and D. Snary. 425 pages. 1987.

Vol. 12: The Cell Surface in Signal Transduction.
Edited by E. Wagner, H. Greppin, and B. Millet. 243 pages. 1987.

Vol. 13: Toxicology of Pesticides: Experimental, Clinical and Regulatory Perspectives.
Edited by L. G. Costa, C. L. Galli, and S. D. Murphy. 320 pages. 1987.

Vol. 14: Genetics of Translation. New Approaches.
Edited by M. F. Tuite, M. Picard, and M. Bolotin-Fukuhara. 524 pages. 1988.

Vol. 15: Photosensitisation. Molecular, Cellular and Medical Aspects.
Edited by G. Moreno, R. H. Pottier, and T. G. Truscott. 521 pages. 1988.

Vol. 16: Membrane Biogenesis. Edited by J. A. F. Op den Kamp. 477 pages. 1988.

Vol. 17: Cell to Cell Signals in Plant, Animal and Microbial Symbiosis.
Edited by S. Scannerini, D. Smith, P. Bonfante-Fasolo, and V. Gianinazzi-Pearson. 414 pages. 1988.

Vol. 18: Plant Cell Biotechnology.
Edited by M. S. S. Pais, F. Mavituna, and J. M. Novais. 500 pages. 1988.

Vol. 19: Modulation of Synaptic Transmission and Plasticity in Nervous Systems.
Edited by G. Hertting and H.-C. Spatz. 457 pages. 1988.

Vol. 20: Amino Acid Availability and Brain Function in Health and Disease.
Edited by G. Huether. 487 pages. 1988.

NATO ASI Series H

Vol. 21: Cellular and Molecular Basis of Synaptic Transmission.
Edited by H. Zimmermann. 547 pages. 1988.

Vol. 22: Neural Development and Regeneration. Cellular and Molecular Aspects.
Edited by A. Gorio, J.R. Perez-Polo, J. de Vellis, and B. Haber. 711 pages. 1988.

Vol. 23: The Semiotics of Cellular Communication in the Immune System.
Edited by E.E. Sercarz, F. Celada, N.A. Mitchison, and T. Tada. 326 pages. 1988.

Vol. 24: Bacteria, Complement and the Phagocytic Cell.
Edited by F.C. Cabello und C. Pruzzo. 372 pages. 1988.

Vol. 25: Nicotinic Acetylcholine Receptors in the Nervous System.
Edited by F. Clementi, C. Gotti, and E. Sher. 424 pages. 1988.

Vol. 26: Cell to Cell Signals in Mammalian Development.
Edited by S.W. de Laat, J.G. Bluemink, and C.L. Mummery. 322 pages. 1989.

Vol. 27: Phytotoxins and Plant Pathogenesis.
Edited by A. Graniti, R.D. Durbin, and A. Ballio. 508 pages. 1989.

Vol. 28: Vascular Wilt Diseases of Plants. Basic Studies and Control.
Edited by E.C. Tjamos and C.H. Beckman. 590 pages. 1989.

Vol. 29: Receptors, Membrane Transport and Signal Transduction.
Edited by A.E. Evangelopoulos, J.P. Changeux, L. Packer, T.G. Sotiroudis, and K.W.A. Wirtz. 387 pages. 1989.

Vol. 30: Effects of Mineral Dusts on Cells.
Edited by B.T. Mossman and R.O. Bégin. 470 pages. 1989.

Vol. 31: Neurobiology of the Inner Retina.
Edited by R. Weiler and N.N. Osborne. 529 pages. 1989.

Vol. 32: Molecular Biology of Neuroreceptors and Ion Channels.
Edited by A. Maelicke. 675 pages. 1989.

Vol. 33: Regulatory Mechanisms of Neuron to Vessel Communication in Brain.
Edited by F. Battaini, S. Govoni, M.S. Magnoni, and M. Trabucchi. 416 pages. 1989.

Vol. 34: Vectors as Tools for the Study of Normal and Abnormal Growth and Differentiation.
Edited by H. Lother, R. Dernick, and W. Ostertag. 477 pages. 1989.

Vol. 35: Cell Separation in Plants: Physiology, Biochemistry and Molecular Biology.
Edited by D.J. Osborne and M.B. Jackson. 449 pages. 1989.

Vol. 36: Signal Molecules in Plants and Plant-Microbe Interactions.
Edited by B.J.J. Lugtenberg. 425 pages. 1989.

Vol. 37: Tin-Based Antitumour Drugs. Edited by M. Gielen. 226 pages. 1990.

Vol. 38: The Molecular Biology of Autoimmune Disease.
Edited by A.G. Demaine, J-P. Banga, and A.M. McGregor. 404 pages. 1990.

Vol. 39: Chemosensory Information Processing. Edited by D. Schild. 403 pages. 1990.